Head, Eye, and Face Personal Protective Equipment

Occupational Safety, Health, and Ergonomics: Theory and Practice

Series Editor: Danuta Koradecka
(Central Institute for Labour Protection – National Research Institute)

This series will contain monographs, references, and professional books on a compendium of knowledge in the interdisciplinary area of environmental engineering, which covers ergonomics and safety and the protection of human health in the working environment. Its aim consists in an interdisciplinary, comprehensive and modern approach to hazards, not only those already present in the working environment, but also those related to the expected changes in new technologies and work organizations. The series aims to acquaint both researchers and practitioners with the latest research in occupational safety and ergonomics. The public, who want to improve their own or their family's safety, and the protection of heath will find it helpful, too. Thus, individual books in this series present both a scientific approach to problems and suggest practical solutions; they are offered in response to the actual needs of companies, enterprises, and institutions.

Individual and Occupational Determinants: Work Ability in People with Health Problems
Joanna Bugajska, Teresa Makowiec-Dąbrowska, Tomasz Kostka

Healthy Worker and Healthy Organization: A Resource-Based Approach
Dorota Żołnierczyk-Zreda

Emotional Labour in Work with Patients and Clients: Effects and Recommendations for Recovery
Dorota Żołnierczyk-Zreda

New Opportunities and Challenges in Occupational Safety and Health Management
Daniel Podgórski

Emerging Chemical Risks in the Work Environment
Małgorzata Pośniak

Visual and Non-Visual Effects of Light: Working Environment and Well-Being
Agnieszka Wolska, Dariusz Sawicki, Małgorzata Tafil-Klawe

Occupational Noise and Workplace Acoustics: Advances in Measurement and Assessment Techniques
Dariusz Pleban

Virtual Reality and Virtual Environments: A Tool for Improving Occupational Safety and Health
Andrzej Grabowski

Head, Eye, and Face Personal Protective Equipment: New Trends, Practice and Applications
Edited by Katarzyna Majchrzycka

Nanoaerosols, Air Filtering and Respiratory Protection: Science and Practice
Katarzyna Majchrzycka

Microbiological Corrosion of Buildings: A Guide to Detection, Health Hazards, and Mitigation
Rafał L. Górny

Respiratory Protection Against Hazardous Biological Agents
Katarzyna Majchrzycka, Justyna Szulc, Małgorzata Okrasa

For more information about this series, please visit: https://www.crcpress.com/Occupational-Safety-Health-and-Ergonomics-Theory-and-Practice/book-series/CRCOSHETP

Head, Eye, and Face Personal Protective Equipment

New Trends, Practice and Applications

Edited by
Katarzyna Majchrzycka

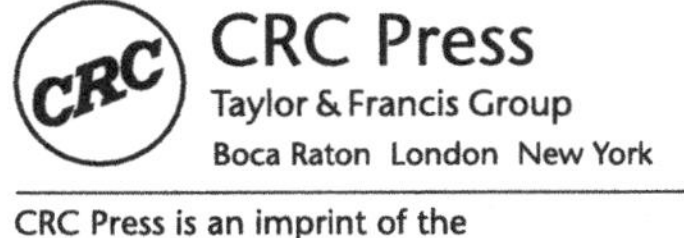
CRC Press
Taylor & Francis Group
Boca Raton London New York

CRC Press is an imprint of the
Taylor & Francis Group, an **informa** business

First edition published 2021
by CRC Press
6000 Broken Sound Parkway NW, Suite 300, Boca Raton, FL 33487-2742

and by CRC Press
2 Park Square, Milton Park, Abingdon, Oxon, OX14 4RN

CRC Press is an imprint of Taylor & Francis Group, LLC

ISBN: 978-0-367-48632-7 (hbk)
ISBN: 978-0-367-52176-9 (pbk)
ISBN: 978-1-003-05680-5 (ebk)

Typeset in Times
by Lumina Datamatics Limited

Contents

Editor

Katarzyna Majchrzycka, PhD (Eng.) DSc (Eng.), is the head of the Department of Personal Protective Equipment (PPE) in Central Institute for Labour Protection – National Research Institute. She carries out research in the area of protective and utility parameters of personal protective equipment, proper selection and development of new solutions and manufacturing technologies. Her primary research focuses on the filtering materials used to protect against bioaerosols. She has participated in several research projects and is the author or co-author of 15 patents and more than 120 publications. She is a laureate of numerous awards in the fields of innovation, scientific research and new technologies in developing advanced PPE.

Contributors

Krzysztof Baszczyński, PhD (Eng.), DSc (Eng.), is a manager of the Laboratory of Safety Helmets and Equipment Protecting Against Falls from a Height in the Department of Personal Protective Equipment of Central Institute for Labour Protection – National Research Institute (CIOP-PIB). He has worked in the institute since 1985. His research interests include test methods and studies of dynamic mechanical quantities associated with the operation of personal protective equipment. His recent publications include articles in the International Journal of Occupational Safety and Ergonomics, Fibres & Textiles in Eastern Europe and Engineering Failure Analysis. He is the author or co-author of 13 patents, 50 publications and 5 monographs. He is also a laureate of numerous awards in the field of test methods and construction of personal equipment protecting against falls from a height and protective helmets.

Marcin Jachowicz, PhD (Eng.), is a researcher at the Department of Personal Protective Equipment of the Central Institute for Labour Protection – National Research Institute (CIOP-PIB). Areas of his scientific interest include head and fall protection equipment, material engineering and thin coating engineering. He is a member of Vertical Group 1 – Head Protection (VG1), and he is the author or co-author of several publications and patents. He is also a laureate of numerous awards in the field of material engineering and personal equipment protecting against falls from a height and protective helmets.

Grzegorz Owczarek, PhD (Eng.), is a physicist and researcher in the field of eye and face protection. He is the head of Eye and Face Protectors Laboratory in the Department of Personal Protective Equipment of Central Institute for Labour Protection – Research National Institute. He has experience in the design of many types of optical protection filters and testing methods of optical and non-optical parameters of these filters.

Joanna Szkudlarek, PhD (Eng.), is a researcher at the Department of Personal Protective Equipment of Central Institute for Labour Protection – National Research Institute (CIOP-PIB), Eye and Face Protectors Laboratory. She conducts research in the field of optical test methods and photometric properties of personal protective equipment including visual functions, which are her main interest. She is a reviewer of the *Fibres & Textiles in Eastern Europe* and *International Journal of Occupational Safety and Ergonomics* (JOSE). She is the author or co-author of 2 patents and more than 20 publications.

Series Editor

Professor Danuta Koradecka, PhD, DMedSc and Director of the Central Institute for Labour Protection – National Research Institute (CIOP-PIB), is a specialist in occupational health. Her research interests include the human health effects of hand-transmitted vibration; ergonomics research on the human body's response to the combined effects of vibration, noise, low temperature and static load; assessment of static and dynamic physical load; development of hygienic standards as well as development and implementation of ergonomic solutions to improve working conditions in accordance with International Labour Organisation (ILO) convention and European Union (EU) directives. She is the author of more than 200 scientific publications and several books on occupational safety and health.

***

The "Occupational Safety, Health, and Ergonomics: Theory and Practice" series of monographs is focused on the challenges of the twenty-first century in this area of knowledge. These challenges address diverse risks in the working environment of chemical (including carcinogens, mutagens, endocrine agents), biological (bacteria, viruses), physical (noise, electromagnetic radiation) and psychophysical (stress) nature. Humans have been in contact with all these risks for thousands of years. Initially, their intensity was lower, but over time it has gradually increased, and now too often exceeds the limits of man's ability to adapt. Moreover, risks to human safety and health, so far assigned to the working environment, are now also increasingly emerging in the living environment. With the globalization of production and merging of labour markets, the practical use of the knowledge on occupational safety, health, and ergonomics should be comparable between countries. The presented series will contribute to this process.

The Central Institute for Labour Protection – National Research Institute, conducting research in the discipline of environmental engineering, in the area of working environment and implementing its results, has summarised the achievements – including its own – in this field from 2011 to 2019. Such work would not be possible without cooperation with scientists from other Polish and foreign institutions as authors or reviewers of this series. I would like to express my gratitude to all of them for their work.

It would not be feasible to publish this series without the professionalism of the specialists from the Publishing Division, the Centre for Scientific Information and Documentation, and the International Cooperation Division of our Institute.

The challenge was also the editorial compilation of the series and ensuring the efficiency of this publishing process, for which I would like to thank the entire editorial team of CRC Press – Taylor & Francis Group.

***

This monograph, published in 2020, is based on the results of a research task carried out within the scope of the second to fourth stage of the National Programme "Improvement of safety and working conditions" partly supported – within the scope of research and development – by the Ministry of Science and Higher Education/National Centre for Research and Development, and within the scope of state services – by the Ministry of Family, Labour and Social Policy. The Central Institute for Labour Protection – National Research Institute is the Programme's main coordinator and contractor.

1 Introduction

Central Institute for Labour Protection – National Research Institute

The working environment as well as everyday living conditions makes it increasingly common to require personal protective equipment to protect the upper part of the head, eyes, and face. There are many reasons for the widespread use of this type of protection. It includes the growing awareness of the threats to the head, eyes, and face. No one is surprised anymore to see a cyclist with a helmet and safety glasses. Technological progress in the construction of helmets, glasses, goggles, as well as other types of head, eye and face protection has made their protective properties increasingly high, while maintaining comfort.

A new trend in the design of head protection is the integration of this type of protection with electronic modules in order to extend their functionality. An example is a helmet with a mounted GPS (*global position system*) module and sensors monitoring basic environmental parameters (e.g., temperature, radiation intensity, etc.). An additional element extending the functionality of traditional eye and face protection equipment are AR (*augmented reality*) modules. Thanks to this type of eye and face protection solutions; they are also used to provide information about the working environment, monitored hazards, etc.

Head, eye, and face protection is most often used in the working environment. Head protection is mainly used in the heavy industry sector (e.g., metallurgy) and in construction and logistics (e.g., warehouses). Eye and face protection equipment is used in virtually all industrial sectors, in medicine and wherever eyes can be exposed to harmful environmental influences.

Due to the wide variety of risks to the eyes and face (harmful optical radiation, mechanical, chemical and biological risks), products to protect against them are also very diverse. There are many types of eye and face protection equipment that are designed for specific applications. Eye and face protection is very often used along with head protection. When wearing protective helmets and eye and face protection at the same time, make sure that it is compatible.

The head is a very sensitive part of the body and due to its location – on the top of the upright silhouette of a man – in the working environment, it is particularly exposed to various risks. The most important of these are

- impacts by dangerous, falling objects,
- impacts on hard, sharp objects in the work area,
- impacts on structural components when arresting a fall from a height,
- transverse compression forces,

- hot factors such as molten metal splashes, open flame, infrared radiation,
- electric shock, and
- the effects of hazardous chemicals in the form of liquids, dust, etc.

The riskiest of the listed factors are mechanical factors, which are clearly confirmed by the data on accident occurrences related to the human head in the working environment in various countries around the world. The scale of the problem can be demonstrated by the following examples of accident data.

According to the National Institute for Occupational Safety and Health (NIOSH) [Industrial Safety & Hygiene News 2017] of 2016, from 2003 to 2010, 2200 construction workers were killed in the United States as a result of accidents related to head injuries. This number represents 25% of all deaths in the construction sector. According to the analysis Tiesman [2011] carried out in the United States between 2003 and 2008 in industrial conditions, about 7300 people died from head injuries, of which 18% were a result of impact with dangerous objects and equipment. The data of the Central Statistical Office [GUS 2018] show that in 2017 in Poland, 5258 incidents related to workers being hit by falling objects were recorded, of which 18 resulted in the death of the injured person and 52 in serious injuries.

The risk of a moving object hitting a worker's head is one of the most serious in many industries. Most often, it occurs in such industries as construction, mining, energy, shipbuilding, warehousing, etc. Depending on the kinetic energy of the moving object, such an impact can result in slight damage to the head surface to the most serious injuries, such as skullbone fracture or direct brain injury.

A similar event is a head-on collision with dangerous objects within the workplace. This can happen with little energy, such as on a low-suspended structural element during normal movement, as well as with high energy, for example, when falling over or falling from a height. Such threats can be found in virtually all industries, particularly in the construction, mining, and energy sectors. Among the mechanical factors posing a threat of head injury are the action of lateral compression forces. Such impacts can cause serious head injuries and are found in industries such as mining, storage management, etc., especially when transporting materials and heavy objects.

Analysing the presented accident events from the point of view of the body injury, it can be concluded that the most frequent effects of mechanical factors on a human head are

- superficial injuries mainly affecting skin,
- damage (most often fractures) of skull bones,
- damage to the skull base and cervical vertebrae, and
- brain damage.

The issue of the medical effects of impacts to the head and modelling of phenomena occurring at that time was reflected in many scientific publications, examples of which are: [Hutchinson 1998; Marjoux 2008; Newman 2000; Shojaati 2003; Tiesman 2011].

A specific nature of work in industrial conditions also poses risks to the worker's head from factors other than mechanical, such as heat. The most important of these include open flame, splashes of molten metal and intense infrared radiation. Such factors can be found in steel mills, foundries, coking plants, during gas and arc welding, etc., and they threaten to cause head burns.

Workstations related to the construction, maintenance and control of electrical installations also pose a risk of electric shock. This risk can also affect a worker's head in some cases and is particularly dangerous due to the current flow path through the human body. In industrial conditions, other dangerous phenomena may threaten the head of a worker, such as contact with aggressive corrosive chemicals in the form of liquids, dusts, etc. Such contact can be very dangerous, both for the scalp and the whole human body.

To sum up the information provided, it is clear that there are many risks to the employee's head in the working environment. These threats can lead to serious damage to human health and, in extreme cases, to death. For many occupations, it is possible to organise work so that the worker is not in direct contact with the hazards, or it is possible to apply collective protective equipment. However, there are also many jobs where the only sensible solution is to use protective helmets. The issues surrounding the use of protective helmets in industrial conditions cover a number of scientific and practical considerations. The most important of these are

- categories of helmets by type and purpose,
- helmet designs and their protective properties,
- new trends in construction and materials used for helmets,
- research on the properties of helmets under different conditions of use,
- rules for selecting helmets for hazards and workplace conditions,
- helmet compatibility with eye and face protectors,
- basic laboratory testing methods for helmets to examine their protective parameters,
- methods of self-assessment of the technical condition of helmets, carried out by users.

These issues are presented in the following monograph in the ensuing chapters.

Another very important element of human protection is eye protection. The eye is one of our most sensitive organs, and the injuries it suffers at work are very common. The National Institute for Occupational Safety and Health (NIOSH) reports that every day, 2000 US workers suffer work-related eye injuries requiring treatment [NIOSH 2013]. Care and protection of the eye is important regardless of the nature of the work: physical or mental. Damage to sight, or in extreme cases loss of vision, results in a reduced quality of life. The natural eye protection apparatus (eyelids, tear gland, conjunctiva, eyebrows and eyelashes) is not always able to provide effective protection, which is particularly true for the working environment. Factors for which the natural eye protection is insufficient include harmful radiation and mechanical hazards, as well as chemical and biological agents. Eye risk factors can also be dangerous to the whole face.

Eye and face risk factors in the working environment can be divided into three basic groups: optical, mechanical, and chemical (including biological agents). In many cases, there are several hazards at the same time.

Optical radiation hazards include radiation from a natural source (sunlight), artificial light sources (e.g., laser radiation) or radiation generated by technological processes (e.g., welding). Harmful radiation concerns the ultraviolet (UV), visible (VIS), and infrared (IR) ranges. Mechanical hazards are mainly chips of solid substances, liquid splashes and dust. Other risks are those arising from chemical and biological agents.

Harmful optical radiation, depending on the wavelength, endangers various parts of the eye. The range of optical radiation that reaches the retina of the eye (and penetrates it directly) is in the range of 380–1400 nm. The radiation range 380–780 nm is visible radiation (VIS), and above 780 nm is infrared radiation (IR). However, a wavelength of up to 400 nm (UV radiation) can be the reason for many adverse changes in the eye, especially as a result of excessive eye exposure.

UV-C radiation (315–400 nm) and UV-B (280–315 nm) can cause inflammatory damage to the cornea; UV-A (180–280 nm) is mainly responsible for cataracts and VIS (400–780 nm) for photochemical and thermal retina damage. Cataracts and corneal burns can be caused by IR-A radiation (780–1400 nm) and corneal haze, cataracts and corneal burns can be caused by IR-B radiation (1400–3000 nm). Far infrared IR-C (from 3000 nm to 1 mm) mainly causes burns to the cornea.

Biological reactions can only be caused by absorbed radiation. There are two types of reactions in biological tissues caused by optical radiation: photochemical and thermal. The effects of exposure to optical radiation depend on the physical parameters of the radiation (wavelength, intensity in relation to individual wavelengths), the amount of the absorbed dose and the optical and biological properties of the exposed tissue: eye, skin, skin phototype, etc. [CIOP-PIB 2003]. Harmful optical radiation can also cause skin burns. This applies both to infrared radiation (thermal exposure) and ultraviolet radiation. The effects of infrared radiation on the skin occur immediately during exposure to this radiation. The effects of ultraviolet radiation, in the form of erythema or burns, occur some time after exposure.

The second group of threats to the eyes and the whole face, after harmful optical radiation, are mechanical factors (chips of solid substances, splashes of liquids or gases under high pressure). Mechanical injuries to eyesight are frequent consequences of accidents at work, resulting from, among other things, non-application of eye and face protection when working with machines or hand tools.

Chemical hazards are also the cause of accidents at work. Injuries caused by chemical agents are due to eye or face contact with the chemical, such as in laboratories, paint shops, on construction sites and during work on chemical or biological waste disposal.

Employers have a particular responsibility to ensure the safety of their employees. Where it is not possible to eliminate, isolate or weaken the effects of eye and face risks, appropriate eye and face protectors should be used. Their choice depends on the circumstances of exposure (nature and extent of hazards), other protective equipment used and the individual requirements of users, taking into account the condition of the eye. The term “eye condition” refers to both health conditions and the temporary state of the eye (fatigue, adaptation conditions, etc.).

Providing adequate eye protection can significantly reduce the severity of accidents. Eye injuries indirectly affect the risk of random events not directly related to sight injuries. An eye injury often means a temporary loss of concentration or balance, with the risk of an unfortunate accident. According to Occupational Health and Safety [OHS 2008], up to 90% of workplace injuries can be eliminated by using appropriate eye and face protection. Providing adequate eye protection can significantly reduce the severity of accidents.

The following monograph presents basic knowledge of the design, testing methods and requirements for protective helmets, as well as eye and face protectors, the principles of their proper selection and examination of their technical condition. Special attention was paid to the characteristics of optical protective filters. The issues concerning eye protection of people suffering from vision dysfunction and the latest trends in the design of protective helmets and eye and face protective equipment are also described.

In addition to the public knowledge, the monograph contains the results of scientific research on protective helmets, and eye, and face protectors. Topics covered are current and technically niche; they result from the authors' professional experience. The utilitarian nature of the study, with a transition from theoretical knowledge (scientific problem) to the application of solutions, distinguishes this monograph from others.

REFERENCES

CIOP-PIB [Central Institute for Labour Protection – National Research Institute]. 2003. www.ciop.pl/CIOPPortalWAR/appmanager/ciop/pl?_nfpb=true&_pageLabel=P30001831335539182278&html_tresc_root_id=23199&html_tresc_id=23216&html_klucz=19558&html_klucz_spis=. (accessed September 19, 2019).

GUS [Główny Urząd Statystyczny]. 2018. Wypadki przy pracy w 2017 r. Warszawa, Gdańsk: GUS.

Hutchinson, J., M. J. Kaiser, and H. M. Lankarani. 1998. The Head Injury Criterion (HIC) functional. *Appl. Math. Comput.* 96:1–16.

Industrial Safety & Hygiene News. 2017. Construction workers at highest risk of traumatic brain injuries. www.ishn.com/articles/106029-construction-workers-at-highest-risk-of-traumatic-brain-injuries. (accessed September 19, 2019).

Marjoux, D., D. Baumgartner, C. Deck, and R. Willinger. 2008. Head injury prediction capability of the HIC, HIP, SIMon and ULP criteria. *Accid. Anal. Prev.* 40:1135–1148.

Newman, J., C. Barr, M. Beusenberg et al. 2000. A new biomechanical assessment of mild traumatic brain injury. Part 2: Results and conclusions. *Proceedings of the 2000 International IRCOBI Conference on the Biomechanics of Impact*, September 20–22, 2000, Montpellier, France.

NIOSH [The National Institute for Occupational Safety and Health]. 2013. Workplace Safety & Health Topic. Eye Safety. www.cdc.gov/niosh/topics/eye/default.html. (accessed September 19, 2019).

OHS [Occupational Health and Safety]. 2008. Workplace Eye Safety. https://ohsonline.com/articles/2008/03/tips-workplace-eye-sefety.aspx. (accessed September 19, 2019).

Shojaati, M. 2003. Correlation between injury risk and impact severity index ASI. *3rd Swiss Transport Research Conference*, Monte Verità/Ascona, March 19–21, 2003.

Tiesman, H. M., S. Konda, and J. L. Bell. 2011. The epidemiology of fatal occupational traumatic brain injury in the U.S. *Am. J. Prev. Med.* 1(41):61–67. doi:10.1016/j.amepre.2011.03.007.

2 Basic Construction of Safety Helmets and Eye and Face Protectors

Central Institute for Labour Protection – National Research Institute

CONTENTS

The construction of the most important eye and face protectors, with regard to their division into transparent elements (optical filters) and non-transparent elements (serving as carriers for the transparent ones) is also discussed in the following chapter. The main risks in the workplace which the given eye and face protectors offer protection against are listed. The materials from which they are made are discussed and the evaluation of their form, with regard to the newest trends such as innovative construction solutions with in-built electronic equipment for additional functionality, is also in the focus. In addition, specific cases of constructions dedicated for persons with vision dysfunction (persons with colour vision deficiencies; persons with implemented intraocular lens – IOLs) are described.

2.1 BASIC CONSTRUCTION OF SAFETY HELMETS

2.1.1 Industrial Safety Helmets

The most commonly used head protective measure in the European Union countries, in industrial conditions, is a helmet with a construction determined, for example, in EN 397:2012+A1:2012 standard [CEN 2012a]. This construction has many attributes in common with helmets characterised in ANSI/ISEA Z89.1-2014 [ANSI/ISEA 2014] ISO 3873 1st Edition 1977 [ISO 1977] and AS/NZS 1801:1997 [AS/NZS 1997] standard. The main tasks of such a helmet include:

- head protection against the impacts from moving, especially falling, dangerous objects,
- protection against hitting a head on the workstation elements.

Additionally, the helmet may protect against:

- lateral compressive forces,
- electric shock hazard,
- molten metal splashes, and
- thermal factors (high and low temperatures, infrared radiation, etc.).

Industrial helmets may also serve as carrier elements for other personal protective equipment, for example, eye and face shields, hearing protectors, respiratory protective equipment, etc.

Regardless of the diversity of safety helmets types, the following common main construction elements may be indicated: shell, harness and headband (see Figure 2.1).

The main tasks of the helmet shell are:

- prevention against the direct contact of a head with a dangerous object, for example, falling, sharp-edged item or a factor, for example, thermal radiation,
- shock absorption that decreases the value of the force acting on the user's head during the impact. Exemplary force transmission to a rigid head form by the helmet hit with a striker with 49 J energy (EN 397:2012+A1:2012 [CEN 2012a]) is presented in the Figure 2.2.

Shock absorption by a helmet, including its shell, depends on the construction, mechanical characteristics, materials used, and thermal conditions in which the helmet is used [Baszczyński 2014b; Gilchrist 1989; Hulme 1996; Forero Rueda 2009].

Depending on the construction solutions, the shell can be equipped with a peak, brim, rain through, ventilation holes, attachment slots for fastening visors and face

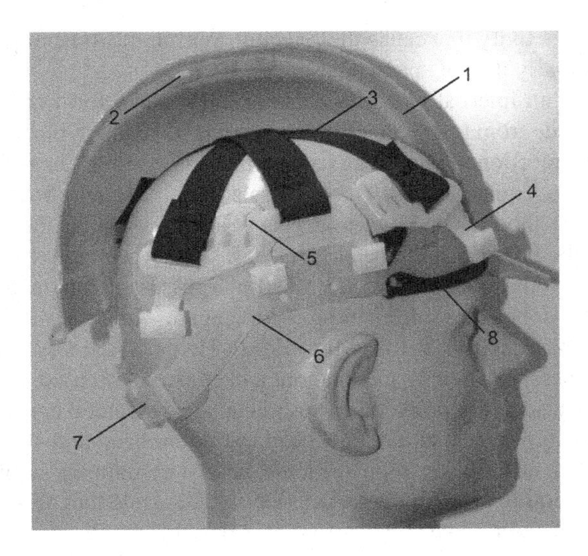

FIGURE 2.1 A typical construction (cross-section) of a safety helmet dedicated for use in industrial conditions. Note: 1 – shell, 2 – ventilation holes, 3 – harness, 4 – harness fixing, 5 – wearing height adjuster, 6 – headband, 7 – length adjuster of the headband, 8 – sweatband.

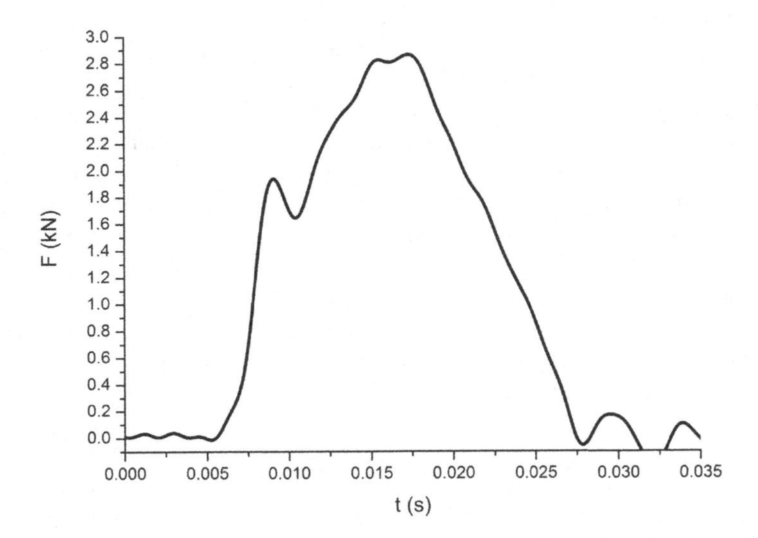

FIGURE 2.2 Force transmission to a headform wearing a safety helmet during the hit with a striker with 49 J energy (testing methods in accordance to EN 397:2012+A1:2012 [CEN 2012a]).

shields and hearing protectors. Helmet shells are usually made of polyethylene, ABS material or various types of laminates, such as glass or Kevlar mats, hardened with synthetic resins. The type of primary product used determines, among others, the permissible helmet rated working duration, due to the plastics' different susceptibility to various factors activity that cause their ageing. Within the most important of these is sunlight, especially its UV spectrum.

The harness is an internal part of a helmet, and it has a form of composition of stripes that are made from textile tapes or from elements made of plastic, for example, polyethylene, by injection method. The harness rests on the user's head. It is connected with the shell by adequate clasps. The harness's main task is to hold a helmet on a head, to absorb the shock of an impact on a shell, and to distribute the forces acting at that moment to the possibly large head surface. It should be noted that a helmet with harness attached to the shell near its brim and not possessing any additional protective padding does not ensure sufficient protection against impacts from the following side directions [Baszczyński 2014a; Mills 1990]. Helmets, the shells of which have an adequate stiffness, ensured by the material and construction used, for example, the prestressed ones, also partly protect the user's head against lateral compressive forces.

A headband is a part that encircles a head at the forehead and skull base height, and, together with a harness, it ensures the helmet remaining stably on a head. The headband is equipped with two regulation mechanisms that allow one to adjust its length and the height of wear. Therefore, it is possible to adjust a helmet to a head girth and increase the helmet stability on the head. Most of the industrial safety helmets have a headband equipped with a sweatband to absorb the sweat produced on the forehead, which increases the comfort during helmet usage.

Most safety helmets are also equipped with a chin strap to prevent a helmet from falling off the head, for example, during bending down, making sudden movements, etc. Usually, anchorages of the chin strap are located on a helmet shell or on a headband. The mechanical strength of the chin strap anchorages slots is adjusted so that on one side, they will not break during a casual use and on the other, they will break during an impact directed from top to bottom of the helmet shell, for example, its peak. Due to such a solution, the user is protected against face traumas in the mandible area. Chin straps may also be equipped with 3- or 4-chin strap anchorages to a shell. This ensures secure helmet attachment to the user's head. It is especially useful in the case when a worker is simultaneously using an individual fall protection equipment and a fall arrest situation is possible [Baszczyński 2018].

Industrial safety helmets may also have additional equipment, such as attachment slots for fastening other personal protective equipment, for example, visors and face shields, miner's lamps, dangerous substances and radiation detectors, location sensors, etc.

2.1.2 High Performance Industrial Helmets

The presented safety helmets, whose construction is accordant with EN 397:2012+A1:2012 standard [CEN 2012a], have a serious fault in that its protective properties are insufficient to protect against side impacts and those from the front and the back. The main cause of this phenomenon is helmet construction, mainly:

- harness attachment in the bottom part of a shell (usually by means of riveting or attachments anchored in the slots),
- small distance between the head and the inner surface of a helmet shell from the front, back and sides.

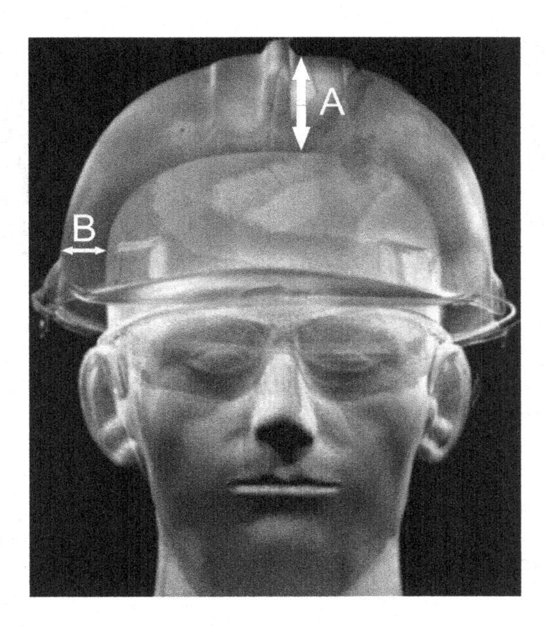

FIGURE 2.3 A comparison of distances between an inner helmet shell surface and a head in the top part (A) and the side part (B) of a helmet.

An exemplary comparison of distances between a shell and a head in the top part (A) and the side part (B) of a helmet is presented in the Figure 2.3. As a result, during a non-central impact on such a helmet, the shell collides with the head with almost no shock absorption, which generates forces of values that threaten the user's health and life. These issues are presented in Baszczyński [2002] and Korycki [2002] papers.

In industrial conditions such as in mining and construction, there are many working environments where the level of mechanical trauma risk, and so side impacts risk, is very high. Therefore, using helmets that meet only the requirements of EN 397:2012+A1:2012 standard [CEN 2012a] is not a solution that would ensure an adequate safety level. The remedy to this problem is using helmets called high performance industrial helmets, which meet EN 14052:2012+A1:2012 standard [CEN 2012b] requirements. An exemplary construction of such a helmet is presented in Figure 2.4.

The main elements of high performance industrial helmet are: shell (1), headband (2) equipped with a length adjuster (3), harness (4) with attachments (5) anchored in a shell, and a protective padding (6). The most important element, compared to construction accordant to EN 397:2012+A1:2012 [CEN 2012a], is protective padding [Shuaeib 2002, 2007; Forero Rueda 2009] located at least in the side part, between a shell and a harness. Usually it is made from polystyrene or polyurethane foam, microporous rubber, PVC and nitrile mix, etc. [Hui 2002; Pinnoji 2010]. The most important feature of these materials is that during the compression (deformation) the force remains approximately stable, which is the essence of shock absorption activity. Shell, harness and padding are responsible for shock absorption

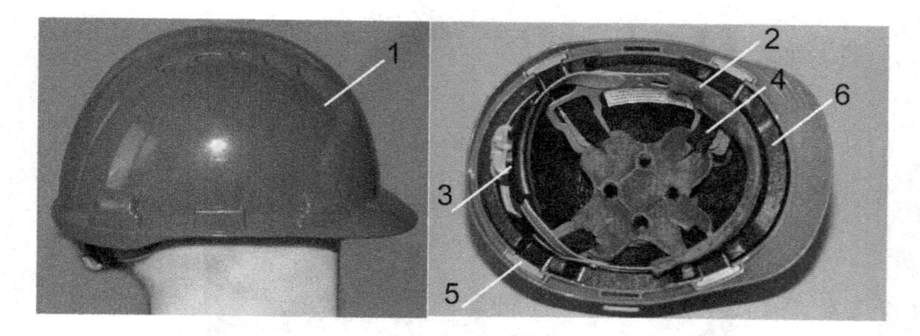

FIGURE 2.4 High performance industrial helmet. Note: 1 – shell, 2 – headband with a sweatband, 3 – length adjuster of the headband, 4 – harness, 5 – attachment point of the harness in a shell slot, 6 – protective padding.

in high performance industrial helmets. Comparing EN 14052:2012+A1:2012 [CEN 2012b] high performance industrial helmets protective properties with helmets accordant to EN 397:2012+A1:2012 [CEN 2012a], the following main advantages can be indicated:

- they ensure a higher level of protection against sharp-edged objects impacts,
- they limit maximum force value that is transmitted to the user's head to a safe value during an impact on a helmet when the kinetic energy of the hitting object is double, and
- they protect a head against both perpendicular impacts (on the shell top point) and impacts from side, front, and back directions.

The disadvantage of high performance industrial helmets, which is a consequence of their construction and the used production materials, is their relatively large mass and their centre of gravity moved upward in comparison to industrial helmets. As a result, it reduces the comfort of using and induces faster fatigue, which causes greater neck muscles stress. A helmet's greater mass can also adversely affect the mechanical stresses that occur during dynamic phenomena, such as fall arrest ensured by individual protective equipment.

2.1.3 Industrial Bump Caps

In many working environments, workers are not exposed to impact from falling dangerous objects, although they are exposed to hitting a head on sharp or hard elements. Such impacts may result only in superficial head damage. In such a situation, it is not necessary to use typical safety helmets or high performance industrial helmets, because the greater protection level they offer is paid with lesser comfort of use. It is mainly due to neck and nape muscles stress caused by greater mass and difficult ventilation in the upper head part. For this reason, in the presented condition the use of the bump caps, which meet the requirements of EN 812:2012 standard [CEN 2012c], is preferred. An example of a typical bump cap construction is showed in Figure 2.5.

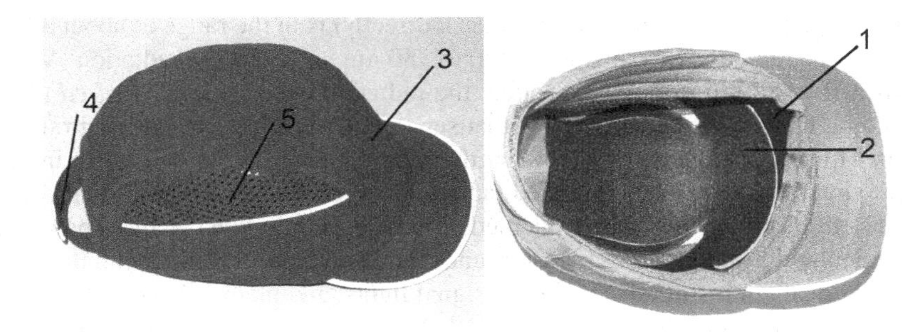

FIGURE 2.5 Industrial bump cap construction. Note: 1 – inner shell, 2 – protective padding, 3 – outer shell, 4 – girth adjuster, 5 – ventilation holes.

The main construction element of the bump cap is the inner shell (1), made by injection of plastic (such as PVC, ABS, polyethylene, etc.). Its main task is to create a barrier preventing direct contact between a user's head and dangerous objects. From the head side, a shell is equipped with protective padding (2), whose task is to absorb the energy of an impact and to create conditions for wearing the helmet comfortably. As a raw material for padding production, various types of soft foam materials are used. The inner shell of a bump cap is covered with fabric formed into a cup shape, whose back is equipped with an adjustable strap that enables adjusting the cap to the user's head girth. In most cases, a bump cap is equipped with ventilation holes, which improves comfort of use.

The bump cap's construction can also be similar to that of helmets accordant to EN 397:2012+A1:2012 [CEN 2012a], which means that they will have a shell, harness, and headband. These elements are, however, lighter and of smaller size.

2.2 BASIC CONSTRUCTION OF EYE AND FACE PROTECTORS

2.2.1 Construction of Eye and Face Protectors and Workplace Hazards

Construction and material used for visors and face shields design are an answer to the need of creating sufficient user protection, corresponding with risks present at a given workstation. The design in terms of the forms of various types of protective equipment, their material structure and complexity of functions is the result of detailed analysis and of knowledge on exposure factors (their types and influence intensity). The knowledge on risks in the working environment and the technological advancement allow for continuous perfection of eye and face protectors construction and for their features and functions adjustment to improve safety and comfort of use.

Eye and face risk factors in the working environment can be divided into three basic groups: optical, mechanical, and chemical (including biological agents).

In many cases, these risks occur at the same time.

Optical risks result from harmful optical radiation, which depends on the wavelength, is a hazard to various eye structures. The range of optical radiation that reaches the retina of the eye (and penetrates it directly) is in the range of about 380–1400 nm. The radiation of range from 380 to 780 nm is the *visible* radiation (VIS), and when the 780 nm value is exceeded, the radiation changes into *infrared* (IR). It should be noted that the boundaries that separate the visible and infrared ranges overlap. This means that the wave of length 780 nm is considered to be both in the visible radiation range and the infrared radiation range. Visible radiation range (i.e., 380–780 nm) is widely adopted in methods used for quantitative calculations (such as for calculating the luminous transmittance value of light that transmits through optical filters, the transmittance factor of signal lights, and many others).

In the classification of harmful effects that accompany eye excessive exposure to optical radiation, the boundary between the *ultraviolet* radiation (UV) and the visible radiation is set by a wavelength of 400 nm. High-energy optical radiation for the wavelength range that directly reaches the retina of the eye, as well as outside this range, can cause many unfavorable changes in the eye. UV-C radiation (i.e., 180–280 nm) and UV-B radiation (i.e., 280–315 nm) can cause inflammatory corneal damage. UV-A radiation (i.e., 180–280 nm) is responsible mainly for cataracts, while VIS radiation (i.e., 400–780 nm) for photochemical and thermal retinal damage. Cataracts and corneal burns can be caused by IR-A radiation (780–1400 nm) and corneal haze and by IR-B radiation (1400–3000 nm). Far infrared IR-C (i.e., 3000 nm–1 mm) causes mainly corneal burns.

Each of the above-mentioned effects of excessive exposure to optical radiation can be of lighter or more severe consequences. Biological reactions can only be caused by absorbed radiation. There are two types of reactions in biological tissues caused by optical radiation: photochemical and thermal. The effects of eye exposure to optical radiation depend mainly on its physical parameters (wavelength, radiation intensity for respective wavelengths), the absorbed dose amount and the biological and optical properties of the exposed tissue: eye, skin, skin phototype, etc.

Harmful optical radiation can also cause skin burns. This applies both to infrared radiation (thermal impact) and to ultraviolet radiation. The effects of infrared radiation acting on the skin are instantly manifested whereas the effects of ultraviolet radiation, in the form of erythema or burns, manifest themselves some time after the exposure.

The second group of risk factors causing impaired vision or damage leading to vision defect or loss, after the harmful optical radiation, are mechanical factors. Mechanical injuries to vision are frequent consequences of accidents at work, resulting due to the lack of using eye and face protectors when working with machines and hand tools. They are caused by chips of solid substances, liquid splashes or gases under high pressure. Frequent incidents caused by mechanical factors include sprinkling eyes with metal shavings (locksmith, turner, car mechanic) and with wood shavings (carpenter, turner, chippy, installer). Severe traumas may be caused by hot liquids (boiling water, molten metal burn) or exposure to the influence of hot vapour, spark, flame or expanding gas. Mechanical factors can also be a severe hazard to the entire face.

The next group are chemical hazards. Damages caused by chemical agents are caused by eye or face contact with chemicals, which happens in laboratories, during works involving chemical substances spraying (e.g., in paint shops), construction works (e.g., calcium oxide burn) and works involving chemical or biological waste disposal.

2.2.1.1 Types of Eye and Face Protectors

The design of eye and face protectors and materials used in them depends on the equipment type and its purpose. Each type or class of eye and face protectors is defined and described in technical standards, containing requirements and testing methods concerning these types of products. Each type or class name and description may slightly differ, depending upon the state of standards used. However, these differences do not influence the choice of a determined eye protector for a determined workstation use. In the monograph, the term of eye and face protectors (or eye protectors) will be used. It is natural because of the fact that, regardless of the design, all types of these protectors shield eyes and parts of a face. It can be a small eye area in case of typical goggles, an entire face in case of full face shields and also ears, head and neck in case of hoods or face shields equipped with additional shields.

In European standards for personal protective equipment [CEN 1995, 2001], the eye and face protectors are divided into spectacles, goggles, face shields (including mesh face shields), and welding shields. Welding shields include shields, safety helmets, welding hoods and googles. The division of eye and face protectors in European standards is slightly different from the division described in international and American standards. In the international standards (ISO) [cf. ISO 4004, cf. ISO 16321-1, cf. ISO 18526-2, cf. ISO 18526-3], eye and face protectors are divided into three main groups: (1) sun protectors, (2) for professional usage, and (3) for sport usage. Furthermore, the ISO 4007 [cf. ISO 4007] international standard that contains main terms and definitions used for eyes and face protective equipment distinguishes the following types of protectors: eye guard (eye shield), face protector (face guard, face screen, face shield), goggles, hand shield, helmet, protective mask, spectacles, visors, clip ons, and prescription parts. In turn, in the Canadian standard [CSA 2015], seven protection classes are distinguished: (1) spectacles, (2) goggles, (3) welding helmets, (4) welding hand shields, (5) hoods class, (6) face shields, and (7) respirator facepieces.

In Figures 2.6 through 2.8, spectacles, goggles, and face shields are shown.

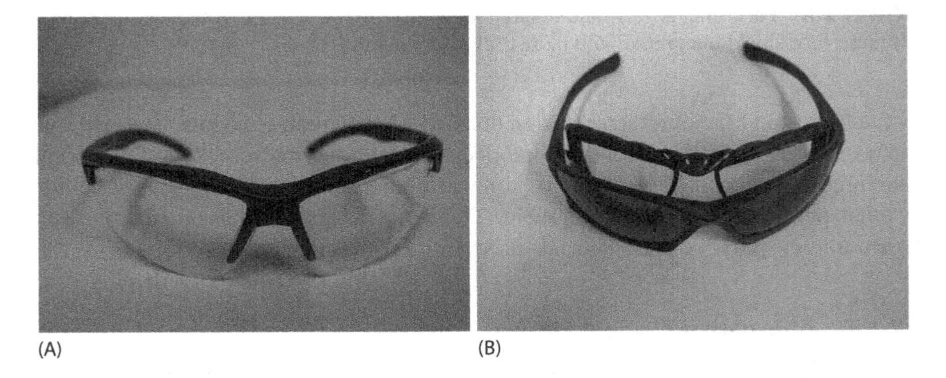

(A) (B)

FIGURE 2.6 Spectacles (A) with transparent protective lens; (B) with filter protecting against sun glare and cartridge protecting against face traumas.

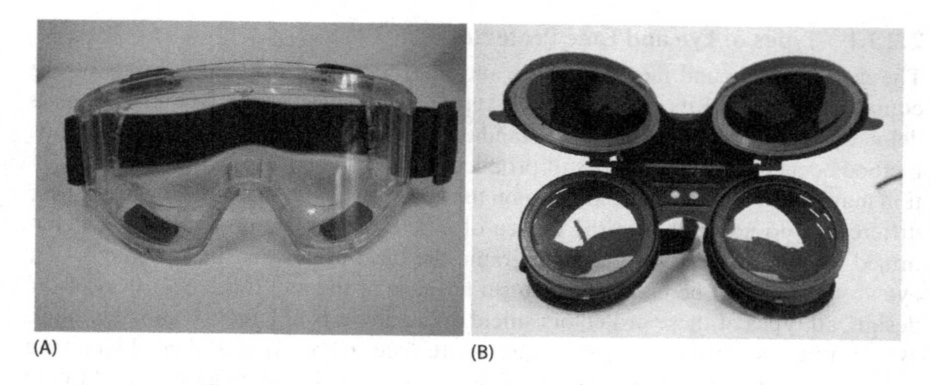

(A) (B)

FIGURE 2.7 Safety goggles. (A) Splash guard goggles; (B) flip-up goggles for welders.

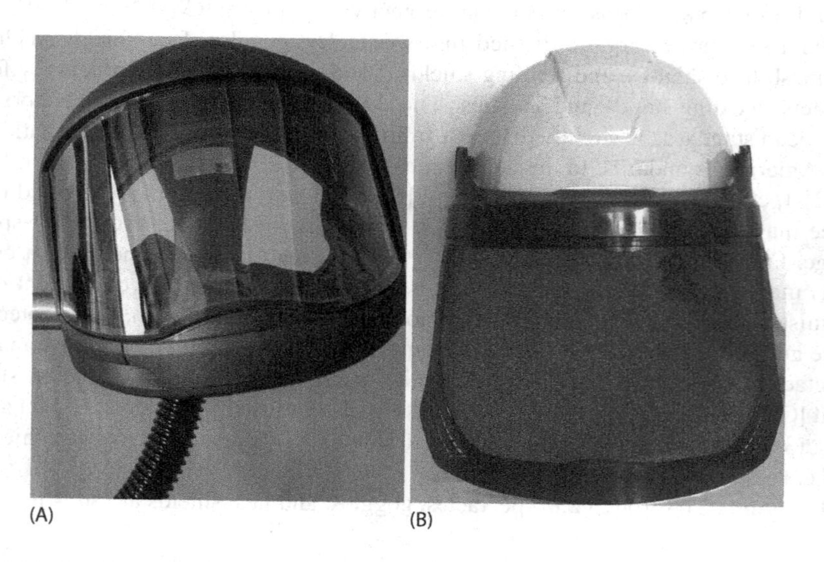

(A) (B)

FIGURE 2.8 Face shields. (A) Face shield integrated with respiratory protective equipment; (B) mesh face shield integrated with industrial safety helmet.

Each eyes and face protector design consists of two main elements. The first one is a transparent element and the second one is everything that serves the function of attaching that element to the head or connecting it with an industrial safety helmet or other personal protective equipment. The basic designs of these elements, with regard to their purpose and materials used for their making, are described below.

2.2.1.2 Transparent Element Design

Transparent elements used in eye and face protectors are usually called lenses, oculars or visors. In common acceptation, "lens" or "protective lenses" mean transparent elements used in spectacles or goggles. The term "visor" is usually used in case of face shields (mainly for those used in respiratory protective devices). Spectacles with corrective effect elements we look through are usually called lenses. It is an analogy to ophthalmic optics terminology. Each protective lens and visor is, in physical terms, also an optical radiation filter. Even if in the initial visual assessment they are completely transparent, their characteristic feature may be to alleviate harmful optical radiation (usually ultraviolet). Figure 2.9 shows a graph presenting harmful optical radiation spectral transmittance characteristic for transparent colourless polycarbonate and inorganic glass.

The spectral transmittance characteristic presented in Figure 2.9 shows that the radiation range of 200–400 nm (UV) is practically completely suppressed by the tested polycarbonate lens. The spectral transmittance factor of harmful optical radiation for the measured ultraviolet radiation range does not exceed the 0.001% value. Visible radiation is transmitted by both lenses on a very high level and it is, for both cases, about 87%.

An example presented in Figure 2.9 shows the physical properties of many polymeric materials which are used for protective lenses and visors design. These materials absorb ultraviolet radiation and therefore become optical filters against this radiation. The group of polymeric materials most often used for protective lenses and visors include polycarbonate (PC), polymethyl methacrylate (PMMA) or cellulose acetate (AC). Protective lenses with corrective effect are usually made of allyl diglycol carbonate (ADC), which is also called CR-39 and Trivex. It is a prepolymer that was developed in 2001 by PPG Industries based on urethane [Users Guide 2006].

In order to consider a protective lens an optical radiation filter, which can be used for eye protection, its properties regarding this area need to be confirmed by laboratory test results. Protective lenses may be divided into those with filtering effect or those without it. The latter are the transparent, colourless optical elements, for

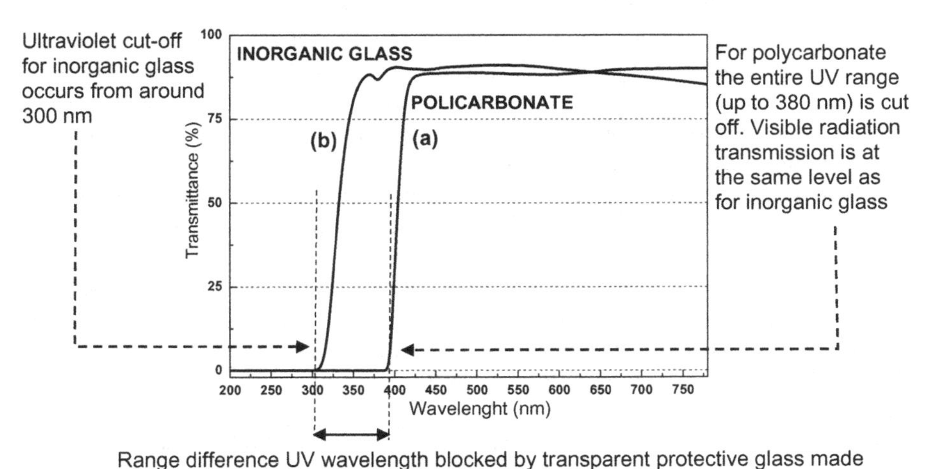

FIGURE 2.9 Harmful optical radiation spectral transmittance characteristic (200–850 nm) for transparent colourless polycarbonate (a) and inorganic glass (b).

which the minimal luminous transmittance value has been determined. According to EN 166 European standard [CEN 2001], the minimal luminous transmittance factor value for protective lenses without filtering effect is 74.4%. In accordance with ISO 16321-1 international standard [cf. ISO 16321-1], the minimal luminous transmittance value for protective lenses of 2 mm and less width is set at 80% level and for protective lenses of 2 mm and more width, it is 75%. Protective lenses with filtering effect, also called optical protective filters, are dedicated for eye protection in hazardous conditions of exposure to harmful optical radiation. Operating principle, design basics and parameters for efficiency assessment of optical protective filters are described in detail in Section 2.2.2.

The "protective lens" term refers both to those elements that are main eye protection and to elements securing optical filters against any damage. Filter security elements are called "cover plate". This is a relatively thin element (from about 0.50 mm), made of transparent colourless plastic (usually polycarbonate), put on a main transparent element that serves eye protection. In order to secure welding filters, installed such as in welding hand shields or in welding safety helmets, cover plates are also used. These are made of inorganic glass of about 2 mm width.

Transparent elements used in eye and face protectors design include meshes. These are used mainly in face shields (mesh face shields) and in spectacles. Mesh spectacles are usually used by miners in underground working conditions, where high risk of chips of solid substances occurs and ensuring lenses' cleanliness at the vision permitting level is very difficult.

All transparent elements may be flat, panoramic or spherical (alike lens used in corrective spectacles). Flat protective lenses may be used in all types of eye and face protectors. Using flat protective lenses in such protectors as face shields may lead to limiting field of vision (lack of possibility of side observation). Despite significantly lesser limitations of field of vision when using a flat protective lens in spectacles or goggles designs, still in these protectors panoramic protective lenses are used. In spectacles, especially in spectacles with corrective effect, spherical protective lenses are also used. These are called just lenses. These lenses are designed like lenses used in corrective spectacles dedicated for medical purposes. These types of designs are further discussed separately in Section 2.2.3, in which the issue of eye protection for persons with vision dysfunctions is discussed. Flat protective lenses are currently used mainly in welding shields and spectacles dedicated for "hot" working environments. This is due to the technology of making welding filters and infrared radiation filters. An example may be technology of making automatic welding filters or interference filters protecting against IR. However, technological advancement within the field of liquid crystal screens and technics of surface modification with vacuous methods allows one to think that in the near future panoramic design of automatic welding filters will also become common [Reviews & Buying Guide 2019]. The same may be true about interference filters and optical protective filters.

Materials of which the protective lenses are made should be characterised with significant mechanical strength and adequate optical parameters. Section 5.2.1 discusses the key issues concerning requirements and testing methods of transparent elements used in eye and face protectors.

2.2.1.3 Design of the Components Used for Holding Transparent Elements

Housings, frames and head harnesses are elements whose purpose is to hold transparent elements of face protectors. The term "frame" refers mainly to spectacles. It is often stated that protective spectacles frames are similar to corrective spectacles in their design. Such a comparison is not entirely correct. The shape, size, and design of corrective spectacles frames are determined not only by its purpose (e.g., housing for installing progressive lenses) but also, and to a great extent, by fashion. In case of protective spectacles frames, it is necessary for frame size to be, regardless of fashion trends, sufficient for protection required by the standards for eye area, which includes also the part of the face at eye height. Thus, protective spectacles frames cannot be too small, and their housing design must protect the side part of the eye area. This can be achieved by using an additional side shield, installed on a temple or provided by temple design itself. Using panoramic lenses in protective spectacles is also a design solution that secures the eye area at the side of the face. Protective spectacles can also be equipped with additional forehead shields. Both side shields and forehead shield makes the design more tight, which decreases the risk of chip or harmful optical radiation entering the eye or eye area. Protective spectacles housing along with lens may form one composite unit. Spectacles of this type are usually made of plastics by injection method.

Typical protective spectacles are manufactured in universal sizes so that different persons, sometimes with different face and head measurements, may use them. Temples are the element that holds spectacles on a head directly. Spectacles temples design is very important because it determines their adjustment and the user's comfort. They need to be characterised by the right dimensions and mechanical strength. Furthermore, the discussed spectacles should alleviate harmful optical radiation that could enter an eye from the side of the face. Under conditions of exposure to harmful optical radiation, its significant dose can enter an eye if spectacles housing is insufficiently tight. Housing side part tightness in relation to optical radiation means that the parameters determining optical radiation alleviation for housing side part cannot be worse than those applied to the lens itself. Mechanical strength must be ensured by selecting adequate materials and by solid connection with that part of the housing to which a protective lens is attached.

Materials used for protective spectacles housing designs are mostly plastics, although there are also housings made of metal (aluminium, stainless steel, titanium, zinc, copper, beryllium, gold, and silver). However, some persons may be allergic to housing metal elements. The most common cause of allergic reactions is nickel presence in the metal alloy. All housing elements adjacent directly to the user's skin must be thus made of materials that do not cause allergic reaction (hypoallergenic materials). These materials include PVC and some silicones, including medical silicone [Morgan 2018].

A separate issue is the design of spectacle housings dedicated for protective lenses with corrective effect installation. This issue is discussed in detail in Section 2.2.3.

However, in working environments, such conditions may occur in which protective spectacles tightness is insufficient. In this case, the use of protective goggles is recommended. The housing design of goggles, which are supposed to fit closely to user's face, provides greater tightness in comparison to spectacles. Goggles housings are mostly made of elastic plastics. Housing shape is cambered so that after right tightening of a headband, the goggles housing fits closely to a face. However, the close fit of goggles housing may result in a foggy protective lens because of sweating. This problem is solved by ventilation systems or by the so-called anti-fog protective lenses. Ventilation systems and anti-fog protective lenses can be used simultaneously. Anti-fog protective lenses are padded from the inside (from eye side) with a layer of a hydrophobic substance preventing the lenses from getting fogged. This layer can be applied permanently on a protective lens. There are also special preparations that users apply on a lens themselves. Efficient rated duration time of these preparations is usually limited. Ventilation systems used in protective goggles housings may differ from each other significantly. The simplest system is a composition of small holes in goggles housing. Ventilation system can also take a form of ventilation ducts. However, the holes in the housing surface may make the goggles design less tight. Properly designed duct ventilation systems allow one to maintain tightness at a high level while at the same time vent the area under the goggles. Examples of goggles ventilation systems that ensure sufficient tightness are shown in Figure 2.10.

The right attachment of goggles to a head is provided by a strap made of elastic material (rubber), which is wrapped around the head. The strap must be equipped with an adjustment system that allows for its fitting to the user's head girth. Its proper attachment and adjustment should make slipping off of goggles during work impossible, but at the same time the strap cannot cause discomfort (due to heavy pressure).

Protective goggles also can be made of metal. This material is used mainly in flip-up goggles for welders. In the goggles of this type, there is an additional flip-up frame in their housing in which welding filters are installed. Observation of the environment with or without the welding filter is possible in these goggles.

Similarly, as in protective spectacles, all housing elements of the goggles that have direct contact with the user's head should be made of safe materials that would not cause allergic reactions.

In the case of face shields, a protective lens is held on a head harness or on a protective helmet. A head harness or a frame that secures a face shield to a safety helmet are made entirely of plastics or plastics in connection with metal elements. A face shield must be equipped with lifting and lowering mechanisms. Its design should enable smooth, effortless shield lifting/lowering, and block its position in

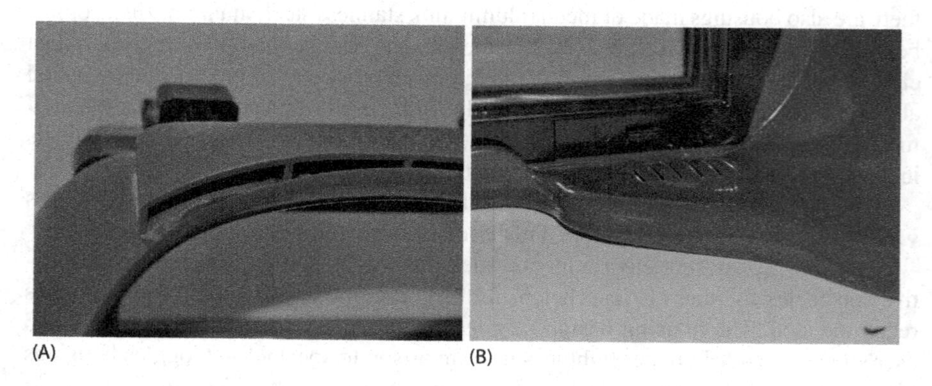

FIGURE 2.10 Goggles ventilation systems: (A) holes in the top part of housing surface; (B) ventilation ducts in the bottom part of housing.

lifted/lowered stages at the same time. Head harness elements that secure it on a head should thus be equipped with an adjustment system that allows for fitting it to a head girth. In some shield designs, especially in those that are characterised with a high mechanical strength (which entails a significant mass increase of the entire shield), elements for head harness height adjustment are also used. The proper head harness adjustment should provide, regardless of the position in which the shield is used (lifted/lowered), that it would not slip off of the user's head in terms of use.

In case of attaching a shield to an industrial safety helmet, the quality of this attachment is determined by both the helmet design and connecting pieces design. Most face shields available on the market, dedicated for using with a helmet, are dedicated for a specific helmet design. Manufacturers of helmets and shields design their products so that specific types of shields fit the specific group of helmets. Less frequently used are designs of on-helmet face shields, which are meant to fit any type of helmet. Analysis of the head harnesses designs that have been present on the market during last three decades shows that there was great advancement made in this field. This concerns both materials themselves and also the adjustment systems. The materials currently used for head harness design are elastic, safe for heath, head-friendly and durable. The ergonomic design of head encircling elements and adjustment systems should also be noted.

The method of attaching a safety helmet on a head is analogous to the face shield case described above. It can be attached on a head harness or on a hard hat. Non-transparent safety helmet material shields the entire face and so observation is possible only through a protective lens of relatively small size. Additionally, a safety helmet may be equipped with ears, neck, and nape shielding elements. Typical use of a safety helmet is to protect the welder's face during welding. Using a welding filter of relatively small size (e.g., 110 × 90 mm) in designs of this type allows one to observe the welded element at the same time. Trends in safety helmets design are currently changing. Traditional safety helmets are characterised by a large protection surface (face and adjacent body parts protection), but also by a small surface of an element through which an observation is possible. Currently, this concerns mainly welding safety helmets, in which an increasingly larger place is occupied by a protective optical filter. There are also such safety helmet designs in which filters are of panoramic shape, which significantly increases the field of vision.

When spectacles, goggles, and safety helmets designs are discussed, it also should be noted that the designers of these protective products ensure that the protectors' appearance corresponds with current fashion trends. Nowadays, eye and face protective equipment should not only guard and be comfortable in use, but also have an appearance appealing to users. It is not only of an aesthetic meaning. The protectors designed in accordance with fashion trends, of unique appearance, are much more likely to be used. No one is surprised anymore that spectacles or goggles decorative designs are different for women and for men. The designers are especially interested in welding safety helmets' decorative designs. Their decorative designs can compete with that of commonly used public space personal protective equipment (e.g., bike helmets, cycle helmets, etc.).

Traditional, hand-held welding shields remain a little behind fashion trends. Their design has stayed nearly unchanged since the last century. Shields made of pressboard, fibre or plastics, among others, are still popular due to their low price. Their design features, which are of direct concern to welders, include mass and perfect balance. Shield handle type and location must be so adjusted that the use of the shield is ergonomic. Unfortunately, some designers do not take this requirement fully into account. Some designs – due to the desire to minimise shield manufacture costs – are non-ergonomic, which becomes strenuous after long-term use. Welding shields may be dedicated for professional use (for occupational welders) or for typically hobby actions. Shields for welding amateurs may have significantly smaller sizes because the risks during short welding with amateur welding equipment are less severe.

Protective hoods are similar in design to safety helmets, although the difference is that the face, head, neck, ears and nape are protected by elastic, inflammable textile material or leather. Protective lenses, usually flat, are set in the hoods on stiff housings or frames. Elasticity provided by hoods facilitates work in places, in which, due to the size limitations, the use of safety helmets is difficult.

The design of elements in which a visor is set in respiratory protective equipment depends on the purpose and design of this equipment. Visors are elements of masks with gas filters, equipment with forced airflow, escape hoods and also protective clothing isolating the entire body. Visors, set in the protective equipment of this type, are mostly anti-fog and have panoramic shape. Due to the specific nature of working environment risks (especially during rescue operations), respiratory protective devices can also be equipped with lens wipers that allow one to clean visor outside surface.

2.2.2 Optical Protective Filters

2.2.2.1 Optical Protective Filters Principle of Operation, Division, and Design

The optical protective filters' main task is to create a barrier against harmful optical radiation. Thus, they must ensure such conditions so that an eye is reached only by radiation in an amount not exceeding MPE – maximum permissible exposure – value [Directive 2006], while simultaneously maintaining light transmission at spectral level that enables to comfortably perform activities at a given workstation. On the other hand, optical protective filters as an element of eye protectors design are a barrier for eyes against mechanical risks, such as impacts and chips of solid substances, particles of molten splashes, splashes of liquid chemical and biological substances, and also gases or vapours that may cause sight organ irritations or traumas. Optical protective filters' principle of operation is illustrated by the schematic in Figure 2.11.

Optical radiation, while passing through any optical element (including any optical protective filter), is subjected to three principal processes, such as transmission, forwardscattering (T), reflection and backscattering (R), and absorption (A). The energy balance basic equation expresses the relationship between transmission, reflection and absorption and it is presented with the following formula:

$$A + R + T = 1, \tag{2.1}$$

where A, R, and T are expressed in relative units.

Optical protective filters are designed for protection against harmful optical radiation emitted both by natural sources (the Sun) and artificial ones (optical radiation artificial sources – laser and non-laser with different spectral ranges).

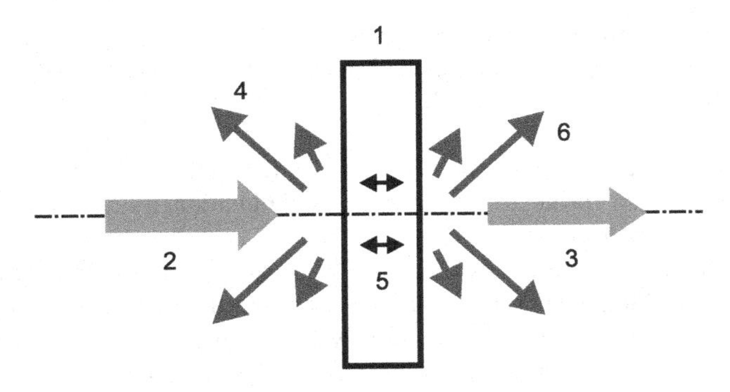

FIGURE 2.11 Optical protective filters principle of operation – a scheme of an optical radiation beam passing through an optical protective filter; (1) – optical protective filter; (2) – optical radiation beam falling on optical protective filter; (3) – optical radiation beam passing through optical protective filter; (4) – backscattering of optical radiation; (5) – part of optical radiation absorbed in a sample; (6) – forwardscattering of optical radiation. (From Owczarek, G. and Jurowski, P. *Prace Instytutu Elektotechniki* (*Works of the Institute of Electrical Engineering*), 255, 201–211, 2012.)

Optical protective filters division refers to optical radiation nature (non-laser and laser radiations), filters purpose (radiation sources and technological processes), and spectral range (ultraviolet, visible radiation, and infrared). This division is illustrated by a schematic shown in Figure 2.12.

Among filters protecting against non-laser radiation, the following may be distinguished:

- filters providing protection against strong sources of ultraviolet radiation (UV protection filters),
- filters providing protection against sun glare caused by visible radiation (mainly solar radiation),

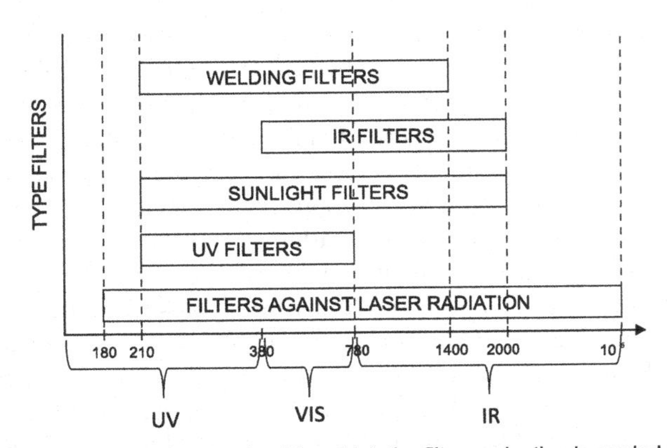

FIGURE 2.12 Optical protective filters division with regard to radiation nature, filter usage and harmful optical radiation range that filters protect against.

- filters used in welding and allied processes,
- filters providing protection against infrared (IR) emitted in "hot" working environments.

In the new ISO international standards [cf. ISO 16321-1, cf. ISO 18526-2, cf. ISO 18526-3), among filters protecting against infrared radiation, a separate category was distinguished: filters for use in glass blowing.

UV protection filters are widely used in industry and medicine. In addition to their capability of absorbing radiation from ultraviolet range, they also shall ensure alleviation (at a sufficient level) of visible radiation, which usually accompanies the ultraviolet radiation. The amount of the accompanying visible radiation depends mainly on ultraviolet radiation source type. Luminous transmittance value in case of this type of filters is therefore correlated with the level at which harmful ultraviolet radiation is blocked. Ultraviolet protection filters [CEN 2002] are used mainly for protection against the radiation of the following sources: low-pressure mercury-vapour lamps (i.e., lamps used for evoking fluorescence or "black light", actinic and bactericidal lamps), average-pressure mercury-vapour lamps (i.e., photochemical lamps), high-pressure mercury-vapour lamps and halogen lamps (i.e., lamps used in solaria).

Filters protecting against sun glare caused by visible radiation are dedicated mainly for protection against solar radiation and sun glare caused by other artificial sources that emit radiation in an open space (e.g., glare caused by vehicle headlights). Due to ultraviolet radiation (of UV-B and UV-A range) occurring in the spectrum of solar radiation, sun glare protection filters shall also provide protection against UV radiation. There are also such sun glare protection filters, for which infrared radiation protection is demanded.

Filters dedicated for use in welding and allied processes provide eye protection against an intense light (glare), as well as against ultraviolet and infrared radiation that are part of the visible radiation spectrum emitted during a certain technological process. The proportion of the amount of radiation emitted in particular spectral ranges (UV, VIS, IR) depends on the welding equipment and technology used. Similarly to UV protection filters, the transmittance factor for this type of filters is correlated with levels at which infrared and ultraviolet radiation are blocked.

Filters serving as protectors against infrared radiation emitted in "hot" working environments provide alleviation of infrared radiation and the accompanying visible radiation. In this type of filters, there is also a rule to correlate luminous transmittance with the level at which infrared radiation should be blocked. The "hot working environment" term refers mainly to metallurgic and foundry industries, where the thermal sources of radiation are molten steel, cast iron, ferrous and non-ferrous metals, slag and molten glass, and they are heated to high temperature walls of a forge, ladle, tank furnace, etc. Other industrial branches may also create such an environment in which the sources are heated to high temperature walls and elements of various types of furnaces (e.g., hardening furnace). Infrared protection filters are also used as face shields in firefighter safety helmets. In the case of filters used in firefighter helmets, protection against ultraviolet radiation is also required.

For filters and complete shields used for protection against laser radiation, the power density (E) and energy density (H) used in energy resistance testing differ much [EN 207:2017]. For filters with the lowest optical density (filters with LB1 code number), dedicated for protection against continuous waves laser (CW Laser) of the range from 180 to 315 nm, the laser radiation power density in energy resistance testing is 0.01 W/m^2. In contrast, for filters with the highest optical density (filters with LB 10 code number) dedicated for protection against pulse laser radiation of 1400 nm to 1000 μm wavelength, with less than 10^{-9} s impulse time (mode-coupled pulse laser), the laser radiation power density in energy resistance testing is as high as 10^{21} W/m^2.

Optical filters can be also divided into passive and active ones. In the case of passive filters, the level of optical radiation transmittance is constant. For their design, inorganic glass or an entire palette of plastics (mainly polycarbonates) is used. The spectral transmittance characteristic modification for "pure" glass or plastic means adding dyes that alter filter darkening or transmission of infrared or ultraviolet radiation to the material the filter is made of (the so-called base material). In addition, passive filter surfaces may be coated with layers modifying the transmission or reflexion of the optical radiation falling on them (mainly the filters' outer surfaces such as anti-reflex coatings) or anti-fog layers (applied to the inner, from the eye side, filter surface). The second category of protective filters are active filters, which are made of materials that allow one to change their optical density in answer to various factors. The activity of such filters may be initiated:

- directly by optical radiation (e.g., photochromic effect),
- indirectly by electromagnetic impulse, which causes changes in the structure of material the filter is made of and which is induced by optical radiation (e.g., liquid crystal screen substances structure changes), and
- polarization effect.

The most known effect that allows for protective filter optical density changes is a photochromic effect. It is induced by optical radiation of ultraviolet range (UV). As a result of ultraviolet radiation falling on the photochromic material, it darkens; thus its optical density increases. The level of visible radiation transmission passing through a photochromic filter is regulated by ultraviolet radiation (UV) intensity changes occurring in the environments where the filters are used. Photochromic effect is commonly used in the design of spectacles lenses dedicated for ophthalmic optics and in sunglasses filters. It is present for a wide range of the ultraviolet spectrum. Photochromic materials used in ophthalmic optics darken under the influence of ultraviolet radiation of UV-B (280–315 nm) and UV-A (315–400 nm) range existing in the atmosphere. The darkening may also be a result of ultraviolet radiation artificial sources that emit radiation of UV-C range (100–280 mn). An example of a solution where a photochromic effect is used also for ultraviolet radiation range emitted from artificial sources (UV-C) is a design of photochromic automatic welding filter [Pościk 2006].

Another factor initiating protective active filters optical properties is light polarization. Optical radiation emitted from most natural and artificial sources is depolarized. A commonly used method of reducing the amount of light (e.g., techniques used in photography) is cutting out a part of visible radiation by using two crossed polarizers. Using light polarization effect in design of a protective active filter with changeable transmission in the visible range is schematically presented in Figure 2.13.

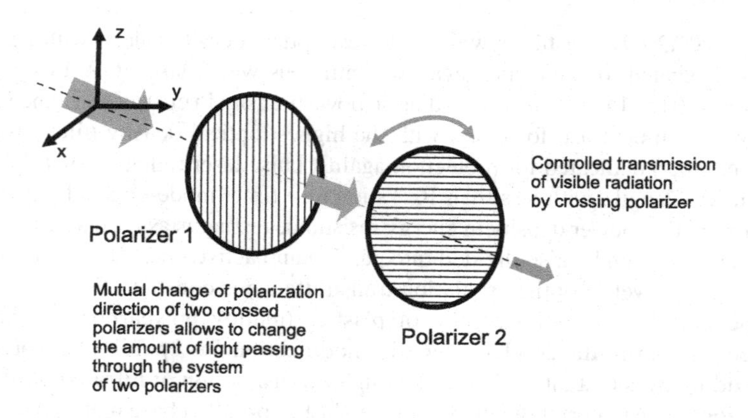

FIGURE 2.13 Polarization filter.

Shown in Figure 2.13 the scheme assumes that the factor responsible for controlling the process of visible radiation transmission is mutual position of two crossed polarizers. This control can be easily achieved both manually and by using an electronic control system equipped with a visible radiation sensor. Polarization effect is also indirectly used in the designs of liquid crystal screens that change the optical density as a result of changes in the direction of nematic of the liquid crystal substance filling this type of a screen. This principle is shown schematically in Figure 2.14.

The described principle is commonly used in automatic welding filter designs. It allows to obtain changes in their optical density, for which radiation transmission level changes from a few percent to thousandths parts of a percent [CEN 2009].

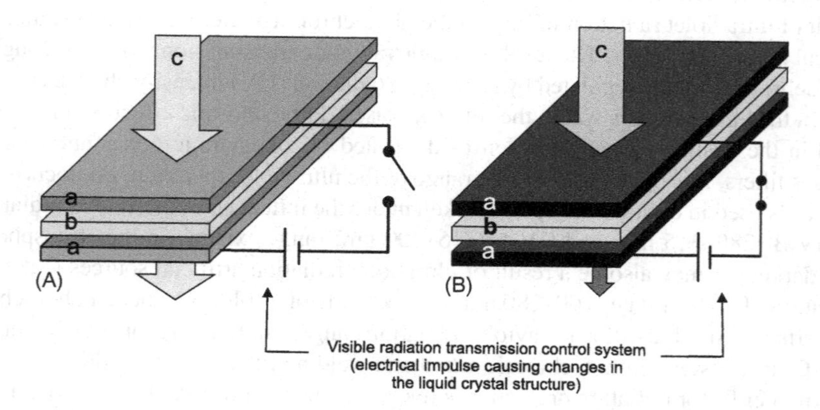

FIGURE 2.14 Liquid crystal screens darkening principle. (A) Screen in light state; (B) screen in dark state. Glass plates – a, liquid crystal – b, optical radiation – c. (From Owczarek, G., Sprawozdanie z projektu 03.8.18. Opracowanie wytycznych dla optymalizacji konstrukcji automatycznych filtrów spawalniczych, uwzględniających rzeczywiste warunki użytkowania, Centralny Instytut Ochrony Pracy – Państwowy Instytut Badawczy, Warszawa, 2001.)

Currently, there are reports about using liquid crystal technology in sunglasses too. They are equipped with powered by battery LCD (*liquid crystal display*) screens and also with a camera analysing the image from the environment. As a result of this analysis, the areas that require local darkening are detected in the filter [Dynamic Eye Inc. 2012]. Active liquid crystal sunglasses design also uses solutions from active shutter glasses, which are dedicated for viewing three-dimensional images (3D). These solutions are at the stage of preparing prototypes and their design, due to the possibility of inducing a darkening only on a small space (about 40 mm^2), do not allow yet to use this technology in protective spectacles.

2.2.2.2 Factors for Examining the Efficiency of Optical Protective Filters

The examination of optical protective filters efficiency is very wide. Factors for examination of protective filters efficiency concern both optical and mechanical properties (e.g., impact resistance, abrasion resistance, aging resistance, etc.). Properties such as mechanical resistance and radiation resistance, against which the protection is to be provided, are extremely important, although without providing a sufficient level of optical radiation filtration they are of no consequence to an optical protective filter design. Optical protective filters' mechanical resistance to impact, as well as their resistance to surface damage due to flying fine particles, their resistance to optical radiation influence, to thermal radiation, etc., can be provided by a proper choice of base and additional protection of filter surface by protective layers. In case of designing optical protective filters, where its main task is eye protection against optical radiation, the most important assumptions are those concerning the transmission and reflection spectral characteristics of the designed filters. These characteristics are a ground for designating factors that determine protective filters' efficiency in terms of protection against optical radiation.

2.2.2.2.1 General Principle of Defining Factors for Optical Protective Filters Efficiency Examination

The wavelengths, which can be taken into account while examining optical protective filters efficiency, are within a range from 200 to 2000 nm. Based on the characteristics of spectral transmittance, the following may be determined:

- spectral transmittance for a given wavelength (τ_λ),
- transmittance factor taking into account the weight functions, which are other spectral distributions (τ_W),
- average spectral transmittance (τ_{avg}), and
- spectral optical density (OD_λ).

Weighted and average ratios and spectral optical density are determined by the following formula:

$$\tau_W = \frac{\int_{\lambda_1}^{\lambda_2} \tau(\lambda) \cdot W(\lambda) d\lambda}{\int_{\lambda_1}^{\lambda_2} W(\lambda) d\lambda} \tag{2.2}$$

where:

τ_W = transmittance taking into account a weight function $W(\lambda)$,
$\tau(\lambda)$ = spectral transmittance,
λ = wavelength (λ_1, λ_2) – wavelengths determining the range, for which transmittance factor is calculated.

$$\tau_{sr} = \frac{\int_{\lambda_1}^{\lambda_2} \tau(\lambda) d\lambda}{N} \tag{2.3}$$

where:

τ_{avg} = average spectral transmittance,
$\tau(\lambda)$ = spectral transmittance factor, and
N = a natural number which corresponds to the value of measurement step interval when measuring spectral characteristic of transmission for wavelength interval between λ_1 and λ_2.

$$OD_\lambda = -\log 10(\tau_\lambda) \tag{2.4}$$

where:

OD_λ = optical density,
τ_λ = spectral transmittance for a given wavelength.

Based on the reflection spectral characteristics, the average spectral reflectance (R_{avg}) is determined according to the following formula:

$$R_{sr} = \frac{\int_{\lambda_1}^{\lambda_2} R(\lambda) d\lambda}{N} \tag{2.5}$$

where:

R_{avg} = average spectral reflectance,
$R(\lambda)$ = spectral reflectance, and
N = a natural number which corresponds to the value of measurement step interval when measuring spectral characteristic of transmission for wavelength interval between λ_1 and λ_2.

In optical filters with a reflection layer, the so-called luminous reflectance is also calculated. It is defined by the following formula [ISO 2001]:

$$\rho_V = \frac{\int_{380}^{780} \rho(\lambda) \cdot V(\lambda) \cdot S(\lambda)_{D65} d\lambda}{\int_{380}^{780} V(\lambda) \cdot S(\lambda)_{D65} d\lambda} \tag{2.6}$$

where:

ρ_V = luminous reflectance,
$\rho(\lambda)$ = spectral reflectance,

$V(\lambda)$ = spectral luminous efficiency function of the average human eye for photopic,
$S(\lambda)_{D65}$ = spectral distribution of radiation of CIE standard illuminant D65, and
λ = wavelength.

2.2.2.2.2 Luminous Transmittance

An example of a transmittance factor that takes weighting functions into account is a luminous transmittance (τ_v), calculated on the base of the formula shown in (2.7). Luminous transmittance is determined in visible radiation range (380–780 nm) in relation to all types of optical protective filters. It shows two spectral distributions: spectral distribution of radiation of illuminant and spectral efficiency function of the average eye.

$$\tau_v = \frac{\int_{380\text{ nm}}^{780\text{ nm}} \tau(\lambda)\cdot V(\lambda)\cdot S(\lambda)\,d\lambda}{\int_{380\text{ nm}}^{780\text{ nm}} V(\lambda)\cdot S_i(\lambda)\,d\lambda}, \tag{2.7}$$

where:
$\tau(\lambda)$ = spectral transmittance,
$S(\lambda)$ = spectral distribution of radiation of illuminant,
$V(\lambda)$ = spectral luminous efficiency function of the average human eye, and
λ = wavelength.

Spectral luminous efficiency function of the average human eye is a relationship determining the amount of energy that reaches an eye in a wavelength function. Vision is different in daylight conditions (photopic) than it is in night-time conditions (scotopic). Figure 2.15 presents spectral luminous efficiency function of the average human eye in daylight conditions ($V(\lambda)_{day}$) [CIE 1926] and in night-time conditions ($V(\lambda)_{night}$) [CIE 1951; Crawford 1949; Wald 1945].

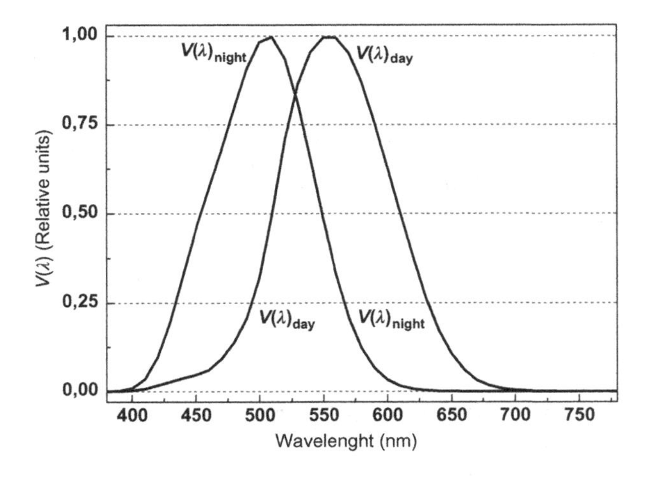

FIGURE 2.15 Spectral luminous efficiency function of the average human eye in daylight conditions ($V(\lambda)_{day}$) (photopic) and in night-time conditions ($V(\lambda)_{night}$) (scotopic).

Distributions presented in Figure 2.15 show that at wavelength range ends in relation to visible radiation (for 380–400 nm and 700–780 nm), vision sensitivity is significantly lower and it drops practically to zero for wavelength end values.

Spectral distribution of outer illumination power is called spectral distribution of radiation of illuminant ($S(\lambda)$). Outer illumination is an important element that is taken into account, for instance, while calculating the value of light transmission through optical elements (e.g., all types of filters used in corrective and protective spectacles, car glasses, etc.). In the calculations, usually two standard spectral distributions of illuminants are used: D65 (natural lighting) or A (artificial lightening). Figure 2.16 presents spectral distributions of radiation of CIE standard illuminants D65 ($S(\lambda)_{D65}$) and A ($S(\lambda)_A$) [ISO 1999].

Spectral distribution of radiation of illuminant influence on luminous transmittance value depends on spectral characteristics (in visible range) of the filter itself. The authors run luminous transmittance tests in relation to welding filters protecting against infrared and colour filters (red, green, and blue filters), also taking into account non-standard, measured under real conditions, spectral distribution of radiation of illuminants (in relation to calculated real spectral distributions of radiation of illuminants, a designation R has been ascribed). For determining luminous transmittance, the seven following spectral radiation distributions of illuminants have been used:

- spectral distribution of radiation of CIE standard illuminant D65 (natural lighting),
- spectral distribution of radiation of CIE standard illuminant A (artificial lighting),
- spectral distribution of radiation of real illuminant R_1 determined in day conditions under cloudy sky,

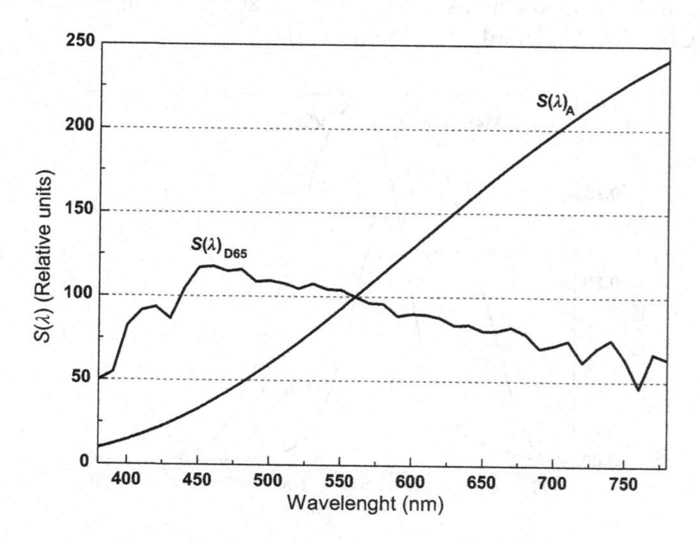

FIGURE 2.16 Spectral distributions of radiation of CIE standard illuminants D65 ($S(\lambda)_{D65}$) and A ($S(\lambda)_{D65}$).

- spectral distribution of radiation of real illuminant R_2 determined in a room lit by artificial incandescent source (halogen lamps),
- spectral distribution of radiation of real illuminant R_3 determined in a room lit by artificial fluorescent source (fluorescent lamps),
- spectral distribution of radiation of real illuminant R_4 determined in a room lit by LED lamps, and
- spectral distribution of radiation of real illuminant R_5 determined in a room lit from a laboratory oven opening with temperature 1000°C.

To determine spectral distribution of radiation of real illuminants (R_2, R_3, R_4), the typical halogen (Osram, 59 W/12 V), fluorescent (Philips TLD 58/840) and LED (SMD 3630, 5,9 W/11 V/350 lm, 3000 K) sources have been used. The spectral distributions of radiation of CIE standard illuminants were determined on the basis of CIE [CIE 1999]. The real spectral distributions were determined on the basis of spectroradiometric measurements. The HR 2000+ (USA) spectroradiometer which has been used was equipped with a fiber and optical radiation diffuser.

In Figures 2.17 through 2.20, spectral distribution of radiation of CIE standard illuminants (D65 and A) are shown in comparison with spectral distributions of radiation of illuminants R_1, R_2, R_3 and R_4 determined in real conditions.

The purpose of determining spectral distribution of radiation of real illuminants was not to compare the spectral characteristics of different types of light but to compare to what extent spectral distribution of radiation of illuminants differ from distributions determined in real conditions, such as a room lit by halogens, fluorescents, LED sources or typical outer conditions under the middle cloudy sky, which translates into transmittance factors of the tested filters' values.

Graphs presented in Figures 2.16 through 2.20 show that the shape of spectral distribution of radiation of illuminants determined in real conditions significantly differs from the shape of spectral distributions of radiation of CIE standard illuminants D65

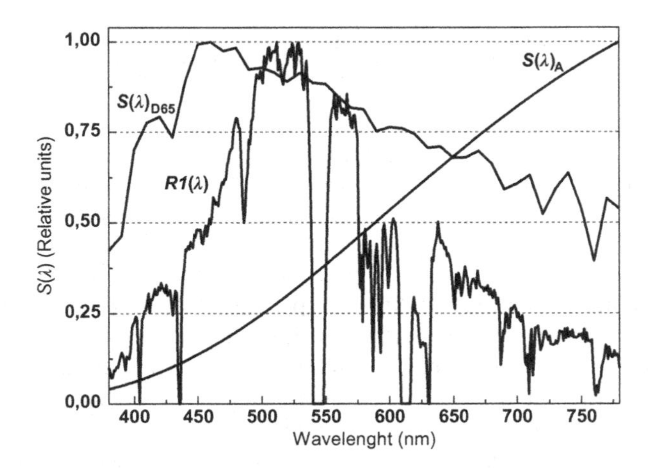

FIGURE 2.17 A comparison of spectral distribution of radiation of CIE standard illuminants A and D65 with spectral distribution of radiation of illuminant R_1 (cloudy sky).

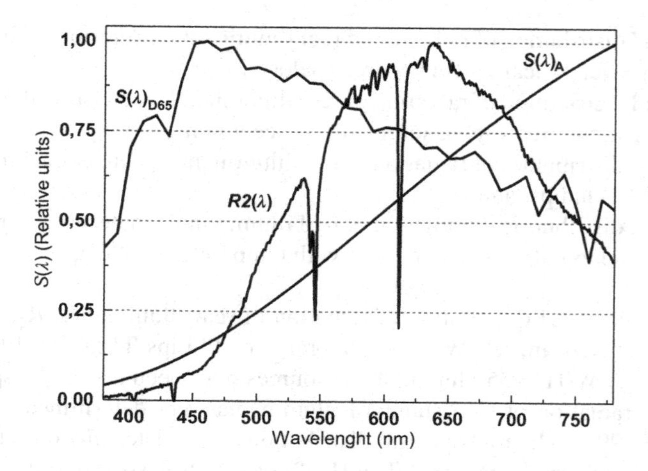

FIGURE 2.18 A comparison of spectral distribution of radiation of CIE standard illuminants A and D65 with spectral distribution of radiation of illuminant R_2 (a room lit by halogen source).

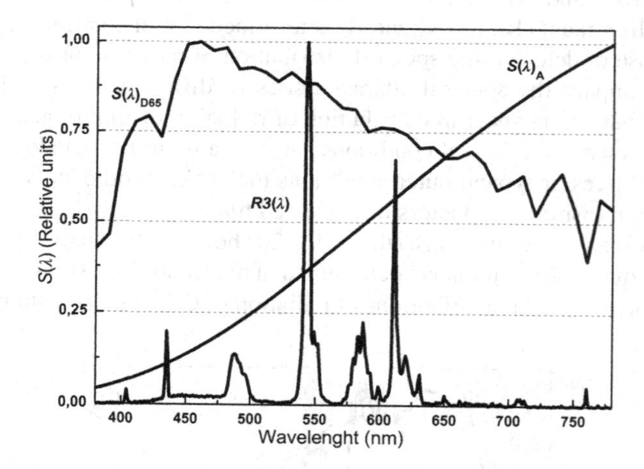

FIGURE 2.19 A comparison of spectral distribution of radiation of CIE standard illuminants A and D65 with spectral distribution of radiation of illuminant R_3 (a room lit by fluorescents).

and A. Of the greatest significance to possible changes in luminous transmittance values, determined with the use of different spectral distributions of radiation of illuminants, may be shifting the distribution maximum in the wavelengths, Of the greatest significance to possible changes in luminous transmittance values, determined with the use of different distributions of illuminants, may be shifting the distribution maximum in the wavelengths, corresponding with maximum eye vision. Analysing only the characteristics of the shape of spectral distribution of radiation of illuminant, it is difficult to state clearly how the change of this distribution may affect the filter luminous transmittance value.

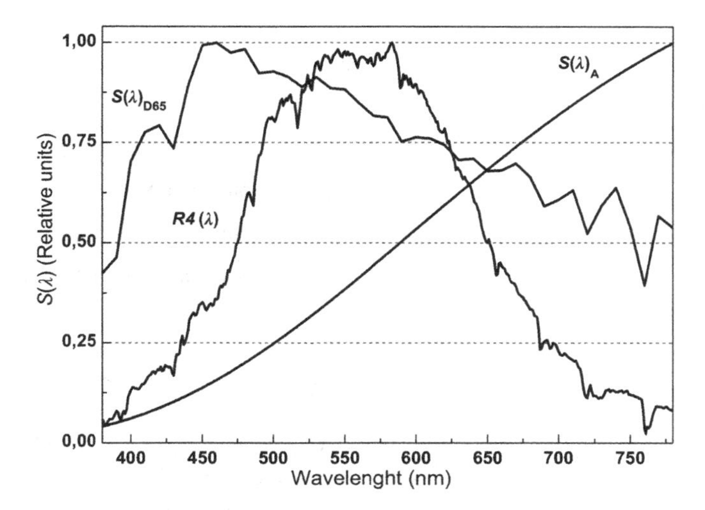

FIGURE 2.20 A comparison of spectral distribution of radiation of CIE standard illuminants A and D65 with spectral distribution of radiation of illuminant R_4 (a room lit by light from LED sources).

In order to achieve that, the values of luminous transmittance for three filter groups (welding, infrared protection, and colour filters) were determined. In each of the tested groups, the lightest and the darkest filters differed significantly in the optical density of visible range. Filters characteristics are presented in Figures 2.21 through 2.26.

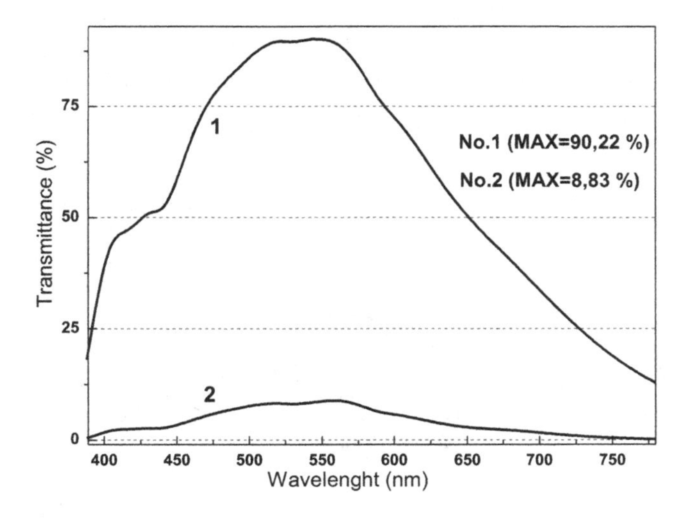

FIGURE 2.21 Spectral transmittance characteristics of filters dedicated for gas welding (samples with the numbers of 1 and 2).

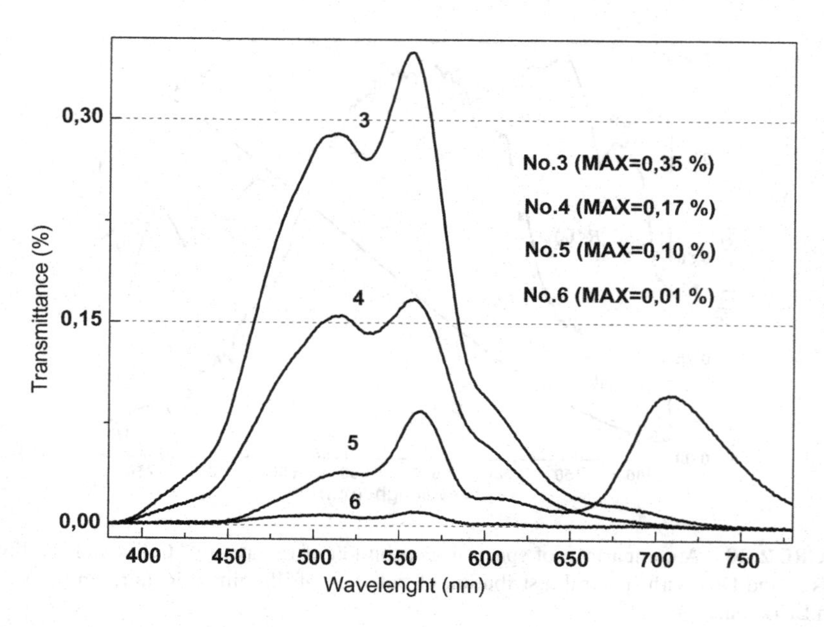

FIGURE 2.22 Spectral transmittance characteristics of filters dedicated for arc welding (samples with the numbers from 3 to 6).

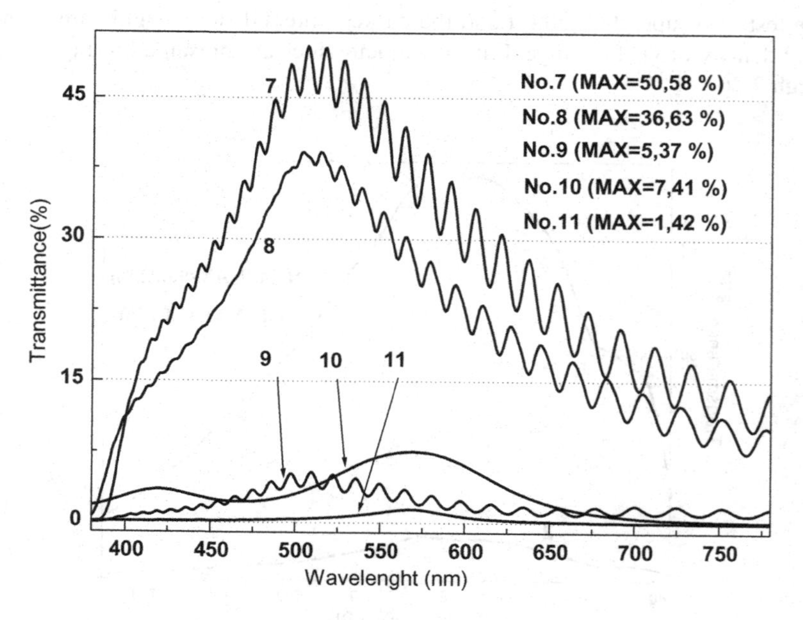

FIGURE 2.23 Spectral transmittance characteristics of filters protecting against infrared (samples with the numbers from 7 to 11).

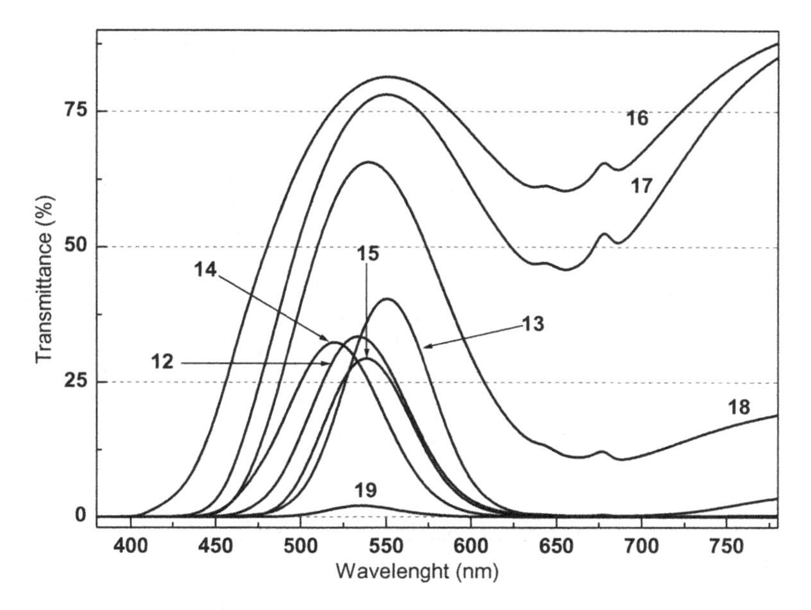

FIGURE 2.24 Spectral transmittance characteristics of green filters (samples with the numbers from 12 to 19).

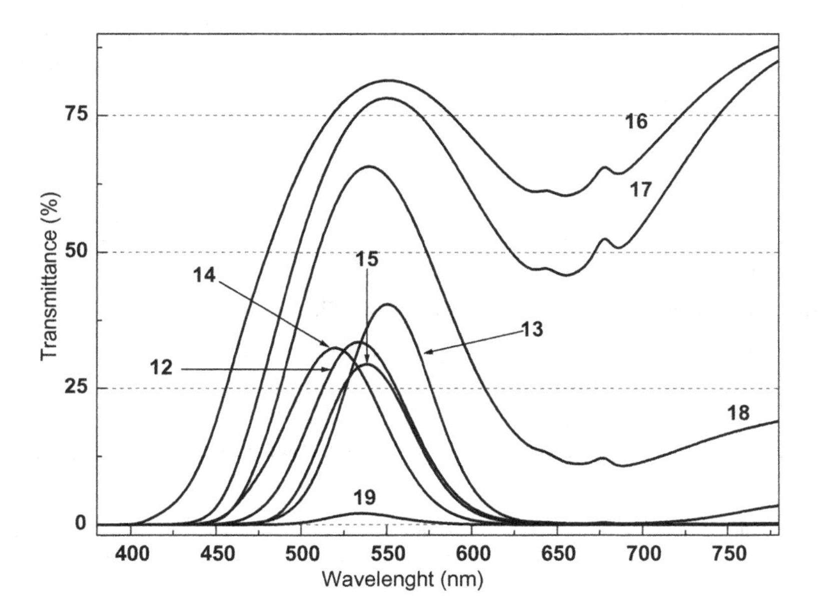

FIGURE 2.25 Spectral transmittance characteristics of blue filters (samples with the numbers from 20 to 26).

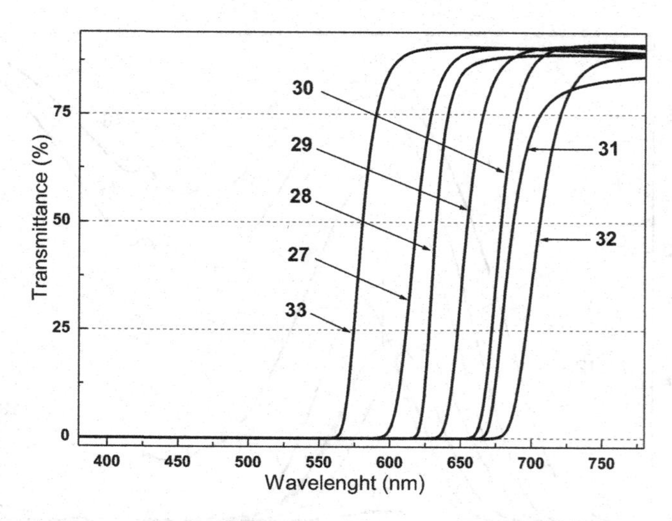

FIGURE 2.26 Spectral transmittance characteristics of red filters (samples with the numbers from 27 to 33).

The results of the determined luminous transmittance values are shown in Tables 2.1 through 2.5.

Analysing the values presented in Tables 2.1 through 2.5, it is easy to see that within all groups of the tested filters (welding filters, infrared protection filters and green, blue and red filters), the greatest influence of distributions of illuminants

TABLE 2.1
Luminous Transmittance Values for Welding Filters

Sample No.	Tested Sample	Luminous Transmittance (%) Standard Illuminants (D65, A)		Real Illuminants (R_1, R_2, R_3, R_4, R_5)				
		$\tau_{vi/day/D65}$	$\tau_{vi/day/A}$	$\tau_{vi/day/R1}$	$\tau_{vi/day/R2}$	$\tau_{vi/day/R3}$	$\tau_{vi/day/R4}$	$\tau_{vi/day/R5}$
1	Filter 2	82.11	79.36	84.46	80.17	83.89	82.78	74.18
2	Filter 5	7.26	6.89	7.63	7.03	7.49	7.38	6.16
3	Filter 8	0.229	0.204	0.253	0.209	0.241	0.233	0.158
4	Filter 9	0.1197	0.1070	0.1320	0.1100	0.1220	0.1210	0.0855
5	Filter 10	0.0432	0.0417	0.0471	0.0433	0.0427	0.0449	0.0366
6	Filter 12	0.00601	0.00532	0.00678	0.00550	0.00581	0.00608	0.00411

TABLE 2.2
Luminous Transmittance Values for Infrared Protection Filters

		Luminous Transmittance (%)						
		Standard Illuminants (D65, A)		Real Illuminants (R_1, R_2, R_3, R_4, R_5)				
Sample No.	Tested Sample	$\tau_{vi/day/D65}$	$\tau_{vi/day/A}$	$\tau_{vi/day/R1}$	$\tau_{vi/day/R2}$	$\tau_{vi/day/R3}$	$\tau_{vi/day/R4}$	$\tau_{vi/day/R5}$
7	Filter 4-3	38.58	36.37	40.46	36.71	38.30	38.59	32.93
8	Filter 4-4	29.56	27.56	31.21	27.76	29.56	29.49	24.58
9	Filter 4-5	3.02	2.70	3.30	2.73	2.83	2.30	2.44
10	Filter 4-5a	2.34	2.37	2.32	2.43	2.58	2.42	2.27
11	Filter 4-7	0.84	0.85	0.87	0.88	0.89	0.89	0.78

TABLE 2.3
Luminous Transmittance Values for Green Filters

		Luminous Transmittance (%)						
		Standard Illuminants (D65, A)		Real Illuminants (R_1, R_2, R_3, R_4, R_5)				
Sample No.	Tested Sample	$\tau_{vi/day/D65}$	$\tau_{vi/day/A}$	$\tau_{vi/day/R1}$	$\tau_{vi/day/R2}$	$\tau_{vi/day/R3}$	$\tau_{vi/day/R4}$	$\tau_{vi/day/R5}$
12	3C-1-1	17.57	14.84	19.66	15.21	20.91	17.92	10.05
13	3C-10-1	21.24	19.73	22.14	20.56	27.34	22.47	15.53
14	3C-11-1	14.29	10.95	27.09	11.06	14.83	13.99	6.19
15	HC3-1-1	14.79	12.76	16.13	13.12	18.79	15.25	8.88
16	HC3-5-1	73.34	73.31	74.25	74.49	76.55	74.96	72.15
17	HC3-6-1	65.73	65.87	66.91	66.96	70.41	67.83	63.29
18	HC3-9-1	46.76	43.76	49.73	44.94	51.95	48.29	36.95
19	HC3-12-1	0.88	0.73	0.96	0.74	1.18	0.89	0.46

occurred for colour filters. Figures 2.27 and 2.28 present a graphic summary of transmittance factor values in relation to the red filter (sample no. 27, filter KC-11-1) and the green filter (sample no. 24, filter C3C-22-1).

2.2.2.2.3 Spectral Transmittance Factors

Spectral transmittance factors and mean spectral transmittance factors of optical protective filters are determined for infrared and ultraviolet radiation ranges. These ranges are taken into account in case of infrared and ultraviolet radiation protection

TABLE 2.4
Luminous Transmittance for Blue Filters

		Luminous Transmittance (%)						
		Standard Illuminants (D65, A)		Real Illuminants (R_1, R_2, R_3, R_4, R_5)				
Sample No.	Tested Sample	$\tau_{vi/day/D65}$	$\tau_{vi/day/A}$	$\tau_{vi/day/R1}$	$\tau_{vi/day/R2}$	$\tau_{vi/day/R3}$	$\tau_{vi/day/R4}$	$\tau_{vi/day/R5}$
20	2-1	51.60	42.84	57.12	43.00	49.87	46.69	31.66
21	C3C-8-1	27.29	19.38	32.32	19.01	23.49	24.77	10.43
22	C3C-9-1	11.69	7.49	14.53	7.16	8.29	10.09	3.12
23	C3C-17-1	67.75	61.12	71.91	61.51	67.24	66.68	52.16
24	C3C-21-1	48.62	37.21	55.86	37.25	47.85	46.10	22.66
25	C3C-22-1	30.24	20.17	36.57	19.46	25.49	26.82	9.08
26	C3C-23-1	67.92	59.08	73.60	59.95	68.53	67.03	46.34

TABLE 2.5
Luminous Transmittance Values for Red Filters

		Luminous Transmittance (%)						
		Standard Illuminants (D65, A)		Real Illuminants (R_1, R_2, R_3, R_4, R_5)				
Sample No.	Tested Sample	$\tau_{vi/day/D65}$	$\tau_{vi/day/A}$	$\tau_{vi/day/R1}$	$\tau_{vi/day/R2}$	$\tau_{vi/day/R3}$	$\tau_{vi/day/R4}$	$\tau_{vi/day/R5}$
27	KC-11-1	7.16	12.81	3.85	11.17	5.38	6.27	22.89
28	KC-13-1	3.43	6.73	2.46	5.72	0.65	2.61	12.90
29	KC-15-1	1.07	2.31	0.76	1.73	0.094	0.64	4.65
30	KC-17-1	0.22	0.53	0.13	0.33	0.014	0.096	1.12
31	KC-18-1	0.12	0.31	0.069	0.19	0.007	0.049	0.66
32	KC-19-1	0.037	0.10	0.018	0.053	0.004	0.012	0.22
33	OC-14-1	25.37	37.65	16.06	36.44	25.49	26.67	55.54

optical filters, filters dedicated for protection against radiation emitted during welding and allied processes and sun protection filters. Spectral transmittance factors of laser radiation protection filters are an exception. For this type of filters, the spectral transmittance factor is determined for a wavelength emitted by a laser, therefore it may as well be a visible radiation.

For optical protective filters dedicated for ultraviolet radiation protection and filters protecting against radiation emitted during welding and allied processes, the spectral transmittance factors are determined for two wavelengths: 313 and 365 nm.

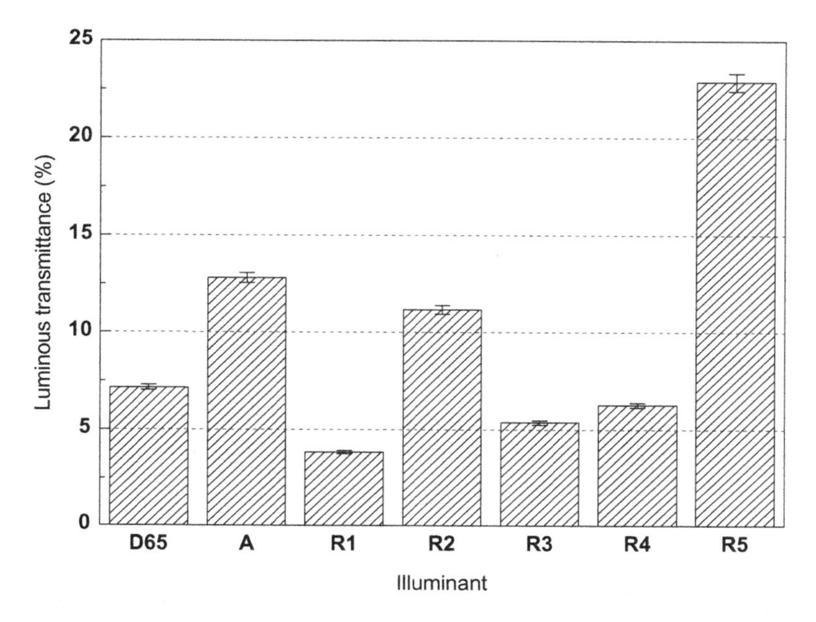

FIGURE 2.27 Luminous transmittance for KC-11-1 red filter (in the figure the 2% measurement uncertainty is marked).

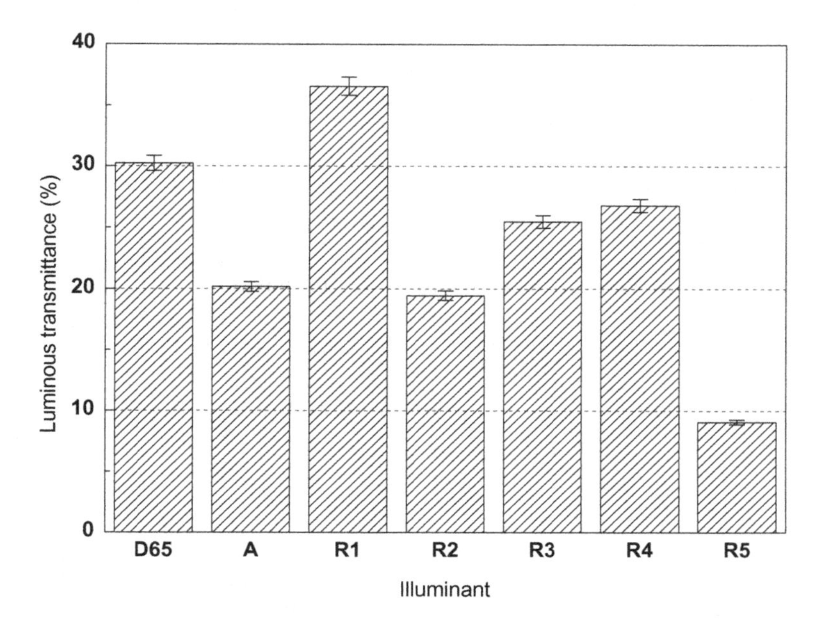

FIGURE 2.28 Luminous transmittance for C3C-22-1 blue filter (in the figure the 2% measurement uncertainty is marked).

For optical protective filters dedicated for infrared radiation protection, mean infrared spectral transmittance factors are determined for two spectral ranges, from 780 to 1400 nm and from 780 to 2000 nm. These factors are determined on the basis of the following formulae:

$$\tau_A = \frac{1}{620 \text{ nm}} \int_{780 \text{ nm}}^{1400 \text{ nm}} \tau(\lambda) \cdot d\lambda \tag{2.8}$$

$$\tau_N = \frac{1}{1220 \text{ nm}} \int_{780 \text{ nm}}^{2000 \text{ nm}} \tau(\lambda) \cdot d\lambda, \tag{2.9}$$

where:

τ_A = mean infrared spectral transmittance factor for 780–1400 nm range,
τ_N = mean infrared spectral transmittance factor for 780–2000 nm range,
$\tau(\lambda)$ = spectral transmittance, and
λ = wavelength.

In the formulae for mean values of transmittance factors, the 1/620 and 1/1220 nm factors (τ_A and τ_N) were introduced due to the measurement step in the spectrophotometric tests value, which is 1 nm.

2.2.2.2.4 *Weighted Transmittance Factors for Sun Protection Filters*

In case of filters protecting against solar radiation, it is especially important to alleviate several UV ranges and infrared present in solar radiation. For this type of filters, the following weighted transmittance factors are determined:

- solar UV radiation transmittance factor (τ_{SUV}),
- solar UV-A radiation transmittance factor (τ_{SUVA}),
- solar UV-B radiation transmittance factor (τ_{SUVB}),
- solar blue light transmittance factor (τ_{Ssb}), and
- solar infrared radiation transmittance factor (τ_{SIR}).

These factors are determined using the following formulae:

$$\tau_{SUV} = 100 \times \frac{\sum_{280}^{380} \tau(\lambda) \cdot E_S(\lambda) \cdot S(\lambda) \cdot \Delta\lambda}{\sum_{280}^{380} E_S(\lambda) \cdot S(\lambda) \cdot \Delta\lambda} = 100 \times \frac{\sum_{280}^{380} \tau(\lambda) \cdot W(\lambda) \cdot \Delta\lambda}{\sum_{280}^{380} W(\lambda) \cdot \Delta\lambda} \tag{2.10}$$

$$\tau_{SUVA} = 100 \times \frac{\sum_{315}^{380} \tau(\lambda) \cdot E_S(\lambda) \cdot S(\lambda) \cdot \Delta\lambda}{\sum_{315}^{380} E_S(\lambda) \cdot S(\lambda) \cdot \Delta\lambda} = 100 \times \frac{\sum_{315}^{380} \tau(\lambda) \cdot W(\lambda) \cdot \Delta\lambda}{\sum_{315}^{380} W(\lambda) \cdot \Delta\lambda} \tag{2.11}$$

$$\tau_{SUVB} = 100 \times \frac{\sum_{280}^{315} \tau(\lambda)\cdot E_S(\lambda)\cdot S(\lambda)\cdot \Delta\lambda}{\sum_{280}^{315} E_S(\lambda)\cdot S(\lambda)\cdot \Delta\lambda} = 100 \times \frac{\sum_{280}^{315} \tau(\lambda)\cdot W(\lambda)\cdot \Delta\lambda}{\sum_{280}^{315} W(\lambda)\cdot \Delta\lambda} \tag{2.12}$$

where:

$\tau(\lambda)$ = spectral transmittance,
$E_S(\lambda)$ = solar spectral power distribution at sea level for air mass 2,
$S(\lambda)$ = relative spectral effectiveness function for ultraviolet radiation,
$W(\lambda) = E_s(\lambda)\cdot S(\lambda)$, and
λ = wavelength in nanometers.

$$\tau_{sb} = 100 \times \frac{\sum_{380}^{500} \tau(\lambda)\cdot E_S(\lambda)\cdot B(\lambda)\cdot \Delta\lambda}{\sum_{380}^{500} E_S(\lambda)\cdot B(\lambda)\cdot \Delta\lambda} = 100 \times \frac{\sum_{380}^{500} \tau(\lambda)\cdot W_B(\lambda)\cdot \Delta\lambda}{\sum_{380}^{500} W_B(\lambda)\cdot \Delta\lambda} \tag{2.13}$$

where:

$\tau(\lambda)$ = spectral transmittance,
$E_S(\lambda)$ = solar spectral power distribution at sea level for air mass 2,
$B(\lambda)$ = blue-light hazard function,
$W(\lambda) = E_s(\lambda)\cdot B(\lambda)$, and
λ = wavelength in nanometers.

$$\tau_{SIR} = 100 \times \frac{\sum_{780}^{2000} \tau(\lambda)\cdot E_S(\lambda)\cdot \Delta\lambda}{\sum_{780}^{2000} E_S(\lambda)\cdot \Delta\lambda} \tag{2.14}$$

where:

$\tau(\lambda)$ = spectral transmittance,
$E_S(\lambda)$ = solar spectral power distribution at sea level for air mass 2, and
λ = wavelength in nanometers.

2.2.2.2.5 Luminous Transmittance in Colour Vision Examination

Colour vision may be particularly important in case of recognising colours of signal lights due to the common practice of replacing traditional fluorescent light sources with LED (*light emitting diode*) type sources. Defining coefficients for quality assessment of colour recognition when looking through optical filters and lenses was preceded with relative spectral distribution of luminance of the signal light analysis and a short analysis of colour recognition mechanism and processes that influence colour vision while looking through colour filters.

2.2.2.2.6 Colour Recognition Mechanism

Whether an eye is able to distinguish a certain wavelength (colour) of the so-called visible range (i.e., 380–780 nm) is an ontogenetic feature and it depends on retina receptor cells' sensitivity. The ability to recognise colours depends on receptor cells (cones), which are located on the retina [Lens 2010]. The process of vision is electrochemical. When the receptor cells (rods or cones) are stimulated by light, the chemical composition of pigment present in these cells is changing. This causes electric current generation to the brain through nerve fibres [Wolska 1998]. Colour vision defects can be divided into congenital and acquired. Congenital defects are classified in accordance to the type of damaged cones and the level of their damage. This classification is shown schematically in the Figure 2.29.

Colour vision defect is usually a congenital defect [Sharpe 1999], genetically determined, inherited recessively, and X chromosome-linked. Due to this reason, these vision defects are much more common in men (about 8% of population) than in women (about 0.5% of population).

The ability to recognise colours may also be disturbed because of using colour filters, i.e., the filters protecting eyes against dangerous optical radiation (used at many workstations) and also the commonly used sunglasses. Colour recognition disturbance while looking through filters may cause a significant inconvenience, for example, during specific technological process observations (e.g., metallurgical processes) or during signal lights observation. The way colour filters alter colour vision can be easily simulated using a computer with graphic processing software, which

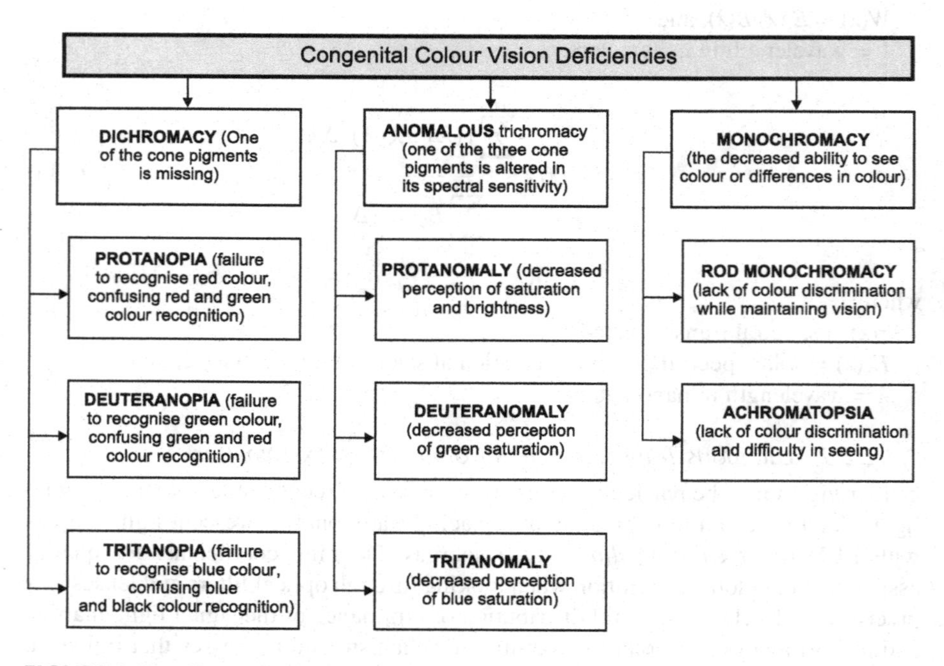

FIGURE 2.29 Congenital colour vision deficiencies.

contains a tool called a "channel mixer". This tool allows for increasing or decreasing saturation of different colours. A spectacular example of allowing for observing an image filtered with channel mixer are Ishihara pseudoisochromatic plates, used for inherent vision defects diagnostics. Figure 2.30 shows an exemplary image form Ishihara plate before and after applying the adjustment of modifying colour palette (exchange of red, green, and blue channel values). The following analysis was performed using GIMP (GNU Image Manipulation Program) graphics software.

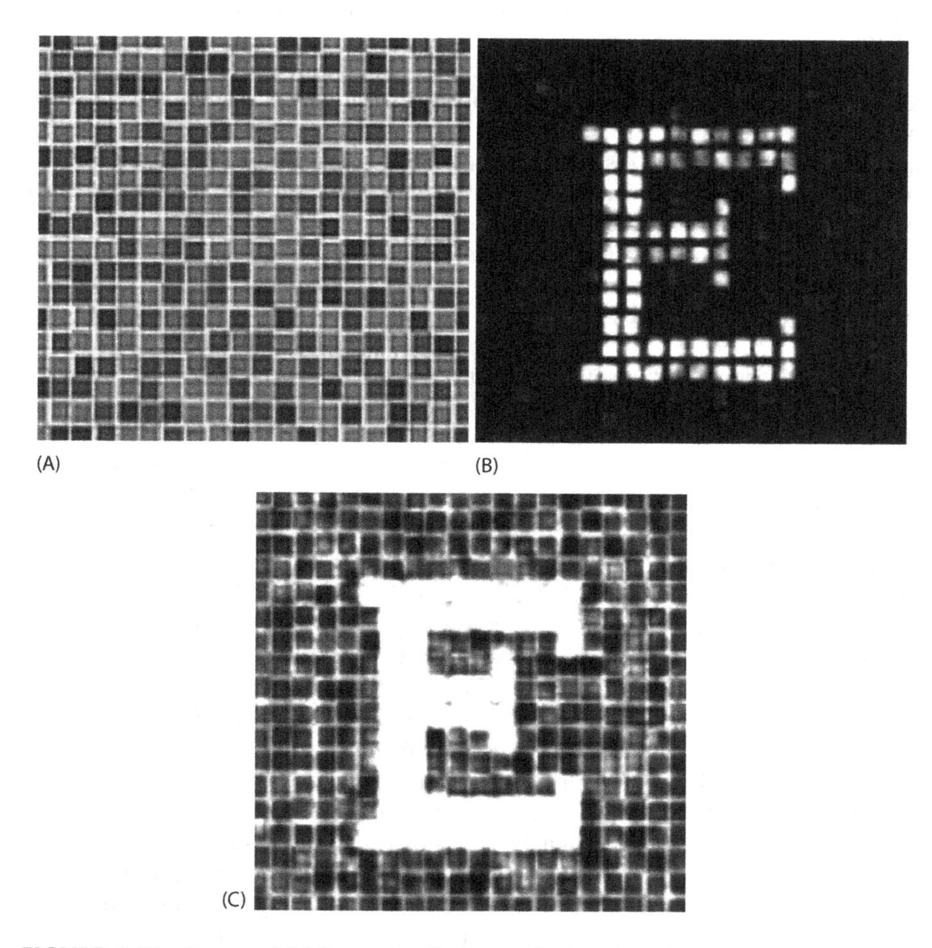

FIGURE 2.30 Image of Ishihara plate before and after the adjustment. (A) Before the adjustment; (B) after the adjustment *(channels: red: 155.6, green: –164.4, blue: –19.3)*; (C) after the adjustment *(channels: red: 191, green: –200, blue: –34.1)*. (From Owczarek, G., Report on statutory task CIOP-PIB. Badania nad zastosowaniem systemów rzeczywistości wzbogaconej dla osób z dysfunkcja narządu wzroku [Research on the application of augmented reality systems for people with visual impairment], Centralny Instytut Ochrony Pracy – Państwowy Instytut Badawczy, Warszawa, 2013.)

Image parameters adjustment for individual vision happens every day, for example, during work with computer screens or watching TV. The basic image parameters (concerning the bit depth in general), such as brightness, contrast and colour saturation are set at a standard level in such a way to provide an optimal vision experience. This term should include such images for which brightness – in the observer assessment – is also optimal. Practically, all commonly used devices with a fundamental function to display images offer the possibility to adjust images for individual users' needs. Correlation of parameters of digital image displayed at screen monitors is also dependent on environmental conditions in which the images are observed. If this takes place in the conditions of an intense light, the images should be of greater contrast and colour saturation.

The issue of adjusting image parameters to outer lightening conditions is well-known to designers of all types of monitors and screens. Modern monitors of computers, smartphones or TV screens can display images of different parameters, according to the designed display modes that adjust image parameters for observation, such as in day conditions (with relatively high outer light illumination) and also in night conditions. Parameters of images displayed by screen monitors in common use are also changed automatically to *adjust* an image to actual outer light conditions.

In the analysis of recognising signal lights colours phenomenon it is necessary to include the following elements:

- human vision and type of outer light,
- signal lights colour,
- spectral distributions of radiation emitted by a signal light source, and
- spectral characteristics of filters through which the signal lights are observed.

Universally present monitors and screens – both in working environment and also in private life – do not leave any place for doubts that in a public space which we live in, the dominating perception is the one registered by an eye channel. Within this space, signal lights are an important element where it is used in road, air and water traffic regulation systems and in widely understood warning and alarm systems. Colour, intensity and light characteristics (continuous or pulse modes) are dependent on the type and use of signal lights. Signals emitted by signal lights are often correlated with sound signals. The most well-known signal lights are red, green and yellow traffic lights. These colours shall be unambiguously recognised by road users, especially by drivers, although it is not absolutely necessary because in case of traffic lights, the red light is always located above yellow and green ones. However, the requirements for full colour vision are restrictively executed in case of many occupations in which signal lights are constant elements of the workplace infrastructure (e.g., railway junctions, seaports and river ports, airports) or of fulfilled activities (occupation of a pilot, sailor, and driver).

In many cases, the colour of signal lights should be recognisable even when wearing spectacles with optical protective filters. This concerns the drivers, pilots or sailors who use sunglasses, as well as laser equipment operators, who are obliged to wear special spectacles protecting their eyes against laser radiation. Recognising colours – including these of signal lights – may be disturbed as a result of organ of vision dysfunction or because of looking through colour filters, or by outer lighting.

2.2.2.2.7 Analysis of Relative Spectral Distribution of Luminance of the Signal Light

The type of signal light source is currently of a great importance due to the widespread use of LED type lights. There are four basic colours of signal lights distinguished: red, yellow, green and blue. Spectral distributions in the case of the mentioned colours of signal lights and source types (LED and traditional fluorescents) are used, among others, in the examination of colour recognition when looking through filters used in optical protective filters, for example, in sunglasses. In Figures 2.31 through 2.34, spectral distributions ($E(\lambda)$) of signal lights were presented, irrespective of vision sensitivity. In Figures 2.35 through 2.38, spectral distributions ($E(\lambda) \cdot V(\lambda)_{day}$) of signal lights were presented, weighted by vision sensitivity in daylight conditions, while in Figures 2.39 through 2.42, spectral distributions ($E(\lambda) \cdot V(\lambda)_{night}$) of signal lights were presented, weighted by vision sensitivity in night-time conditions. Each of the spectral distributions $E(\lambda$, ($E(\lambda) \cdot V(\lambda)_{day}$)), ($E(\lambda) \cdot V(\lambda)_{night}$) was determined on the basis of data contained in PN-EN ISO 12311:2014 [ISO 2014] and data concerning spectral distributions of vision sensitivity in day and night conditions [CIE 1951].

On the basis of the graphs of spectral distributions presented above, a wavelength (λ_{max}) was determined, for which there is maximum energy of radiation and *full width at half maximum* (FWHM) of the distribution. These values were determined

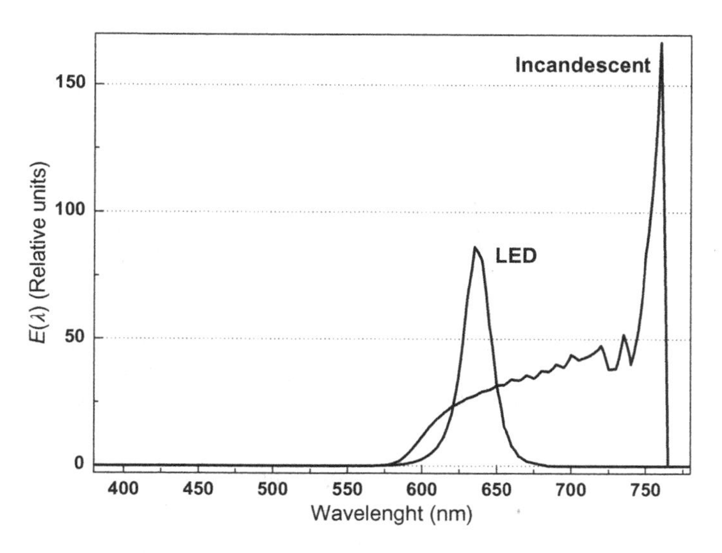

FIGURE 2.31 Spectral distributions ($E(\lambda)$) of signal lights emitted by LED sources and by traditional fluorescent sources for a red colour, irrespective of vision sensitivity.

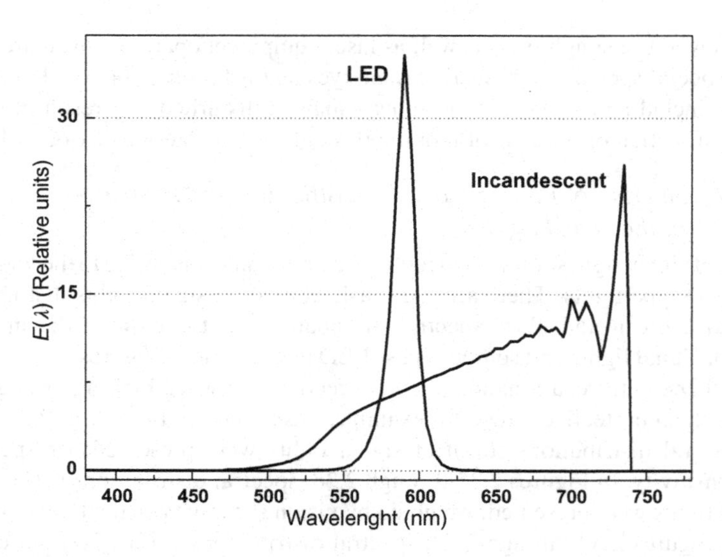

FIGURE 2.32 Spectral distributions ($E(\lambda)$) of signal lights emitted by LED sources and by traditional fluorescent sources for a yellow colour, irrespective of vision sensitivity.

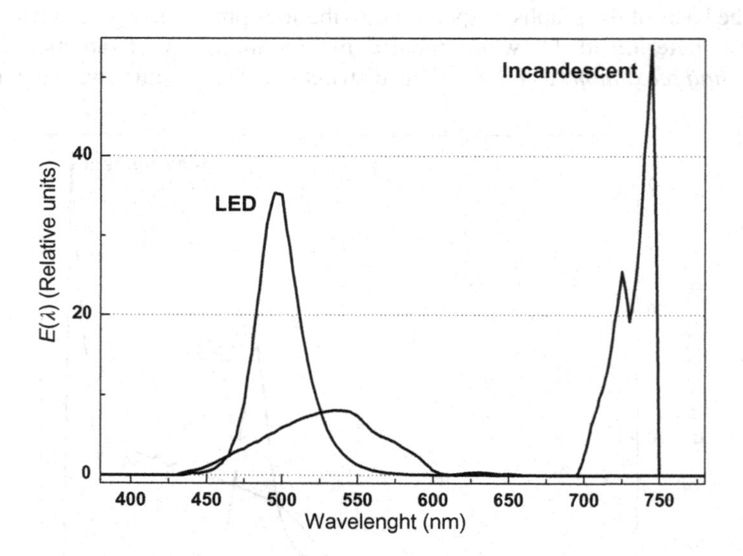

FIGURE 2.33 Spectral distributions ($E(\lambda)$) of signal lights emitted by LED sources and by traditional fluorescent sources for a green colour, irrespective of vision sensitivity.

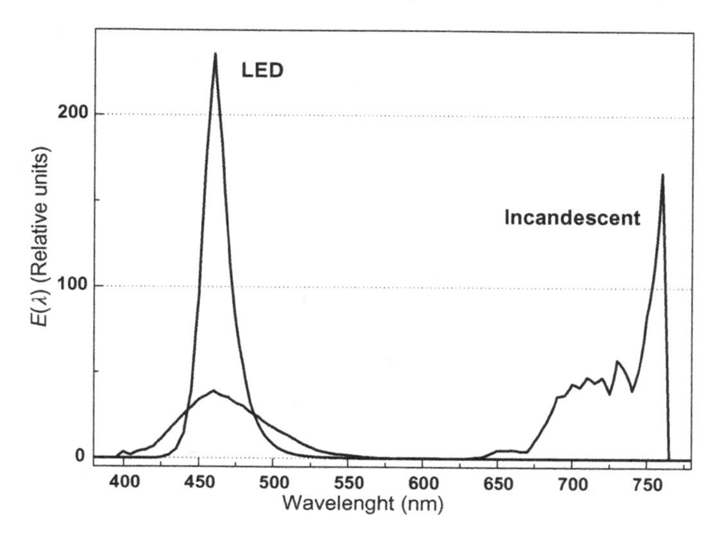

FIGURE 2.34 Spectral distributions ($E(\lambda)$) of signal lights emitted by LED sources and by traditional fluorescent sources for a blue colour, irrespective of vision sensitivity.

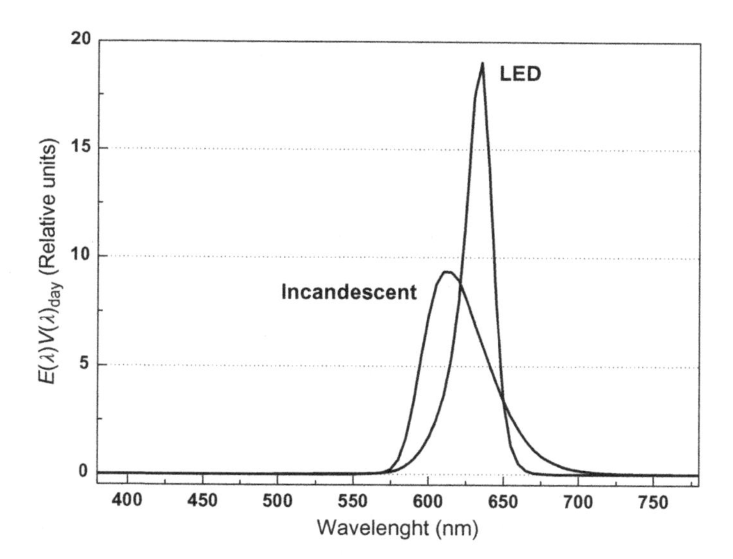

FIGURE 2.35 Spectral distributions ($E(\lambda) \cdot V(\lambda)_{day}$) of signal lights emitted by LED sources and by traditional fluorescent sources for a red colour, weighted by vision sensitivity in daylight conditions.

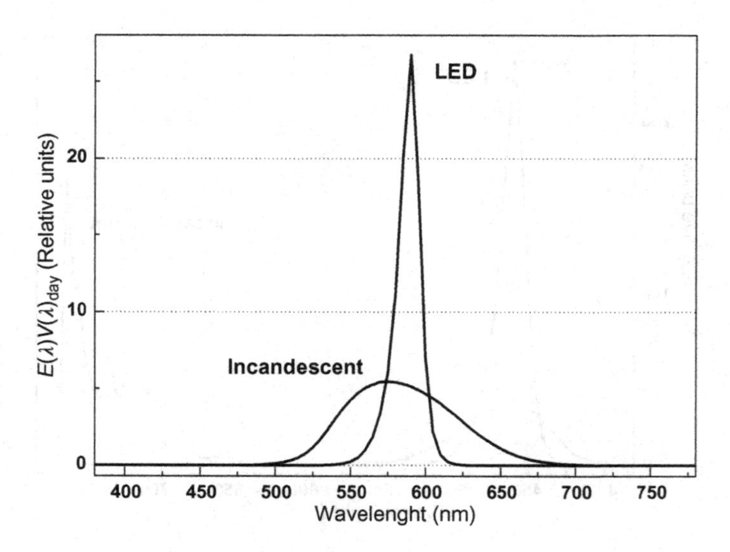

FIGURE 2.36 Spectral distributions ($E(\lambda) \cdot V(\lambda)_{day}$) of signal lights emitted by LED sources and by traditional fluorescent sources for a yellow colour, weighted by vision sensitivity in daylight conditions.

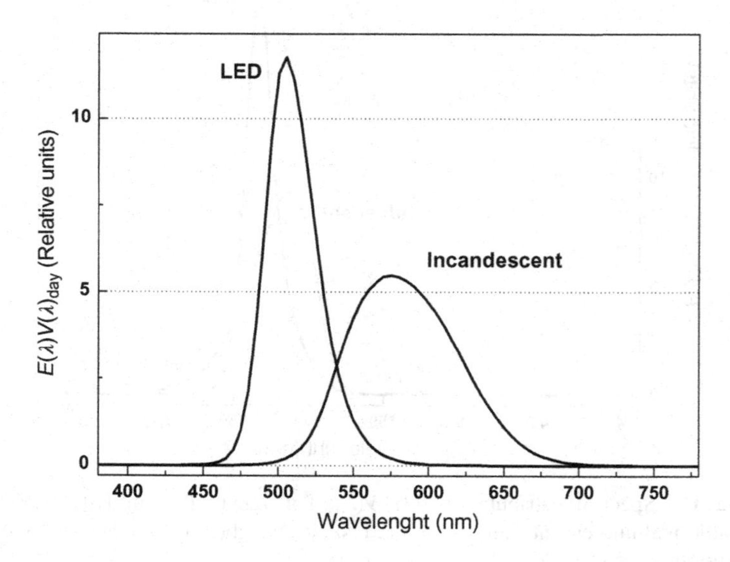

FIGURE 2.37 Spectral distributions ($E(\lambda) \cdot V(\lambda)_{day}$) of signal lights emitted by LED sources and by traditional fluorescent sources for a green colour, weighted by vision sensitivity in daylight conditions.

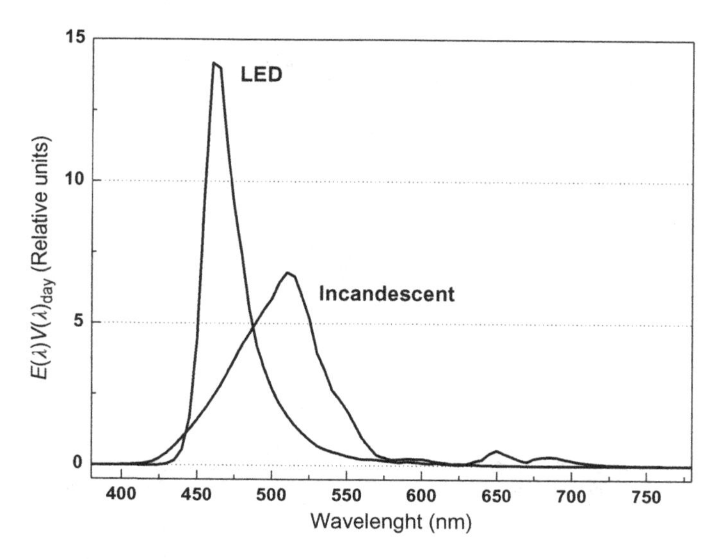

FIGURE 2.38 Spectral distributions ($E(\lambda) \cdot V(\lambda)_{day}$) of signal lights emitted by LED sources and by traditional fluorescent sources for a blue colour, weighted by vision sensitivity in daylight conditions.

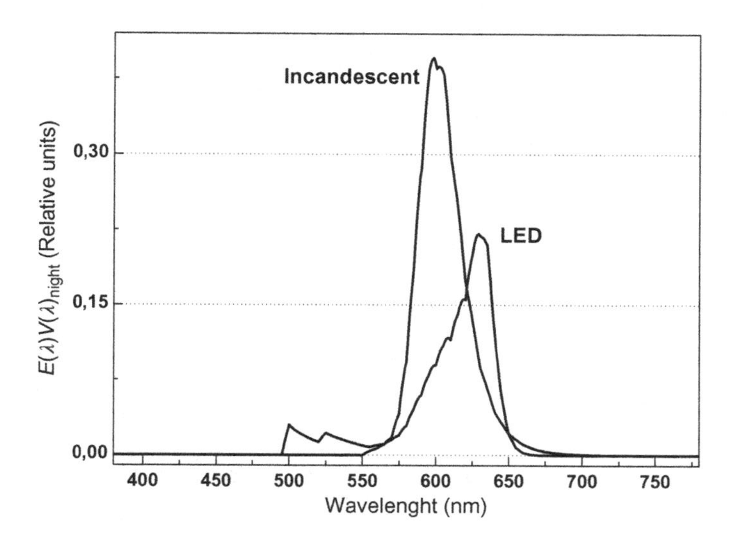

FIGURE 2.39 Spectral distributions ($E(\lambda) \cdot V(\lambda)_{night}$) of signal lights emitted by LED sources and by traditional fluorescent sources for a red colour, weighted by vision sensitivity in night-time conditions.

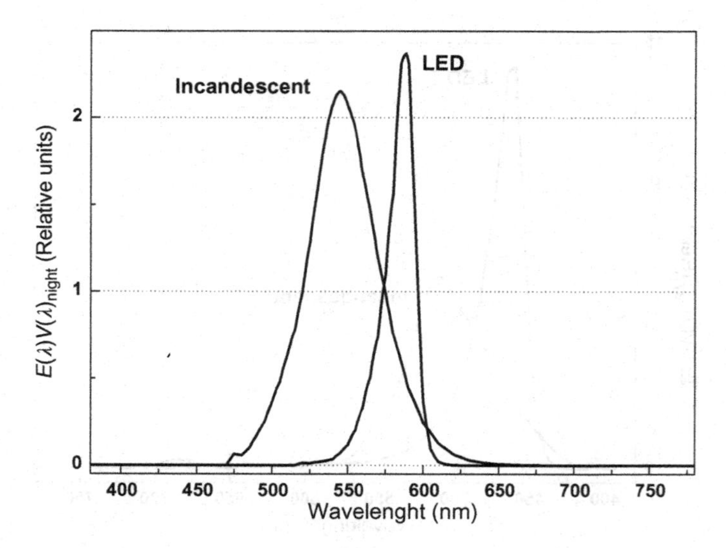

FIGURE 2.40 Spectral distributions ($E(\lambda) \cdot V(\lambda)_{night}$) of signal lights emitted by LED sources and by traditional fluorescent sources for a yellow colour, weighted by vision sensitivity in night-time conditions.

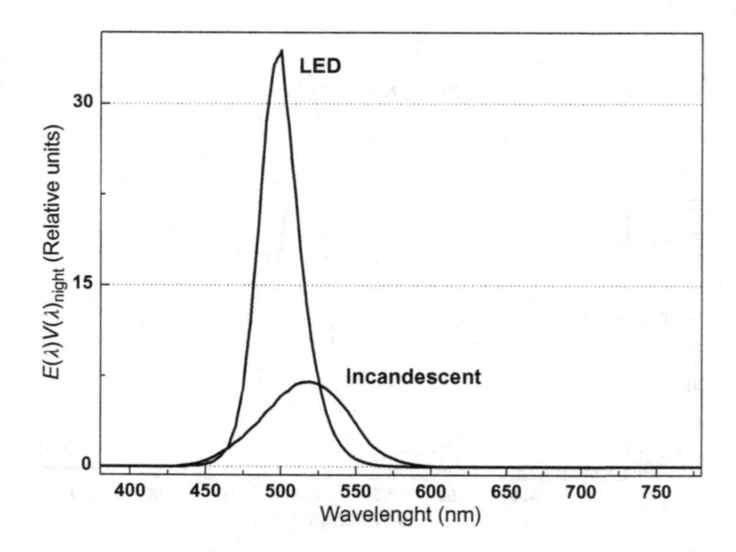

FIGURE 2.41 Spectral distributions ($E(\lambda) \cdot V(\lambda)_{night}$) of signal lights emitted by LED sources and by traditional fluorescent sources for a green colour, weighted by vision sensitivity in night-time conditions.

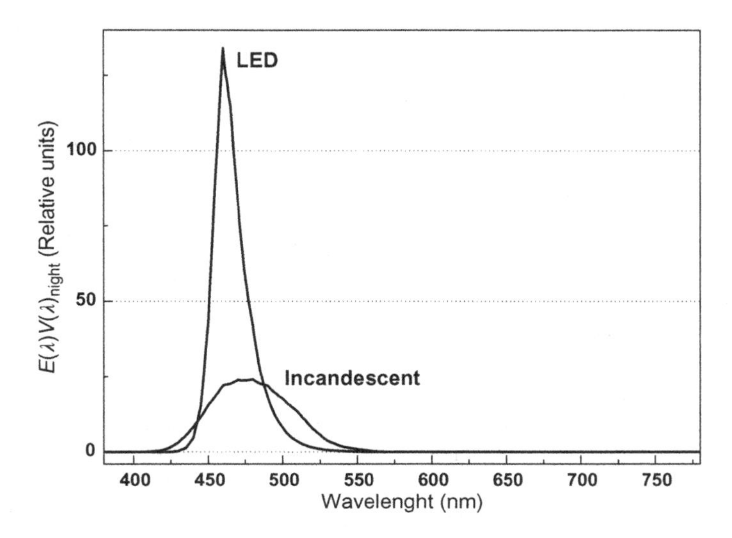

FIGURE 2.42 Spectral distributions ($E(\lambda) \cdot V(\lambda)_{night}$) of signal lights emitted by LED sources and by traditional fluorescent sources for a blue colour, weighted by vision sensitivity in night-time conditions.

by graphs approximation with 1 nm accuracy (with the use of Origin software). The results were presented in Table 2.6.

Results presented in Figure 2.6 show that:

For LED signal lights:

- wavelength (λ_{max}), for which the maximum energy of radiation does not change significantly, regardless of whether a spectral distribution is weighted by vision sensitivity in daylight or night-time conditions, or if the distribution does not take vision into account at all. Maximum shift of λ_{max} value, that is only 6 ± 1 nm, is observed between blue signal light spectral distribution, weighted by vision sensitivity in daylight conditions ($E(\lambda) \cdot V(\lambda)_{day}$) and the distribution for this light, weighted by vision sensitivity distribution in night-time conditions ($E(\lambda) \cdot V(\lambda)_{night}$). For the remaining colours of signal lights, these shifts are: 6 ± 1 nm for red colour, 2 ± 1 nm for yellow colour, and 0 ± 1 nm for blue colour. For all colours the maximum energy of a distribution weighted by vision sensitivity in daylight conditions is the shift in relation to maximum determined in night-time conditions in the longer waves direction, which is obvious due to the fact that there is relatively significant shift between vision sensitivity maximums in daylight conditions (vision maximum for $\lambda = 555 \pm 1$ nm) and in night-time conditions (vision maximum for $\lambda = 510 \pm 1$ nm);
- all analysed spectral distributions have a shape of a relatively narrow peak. These peaks' full width at half maximum is between 16 ± 1 nm (in case of yellow colour distributions) to 40 ± 1 (in case of green signal

TABLE 2.6
Spectral distribution analysis for signal lights

Type of Source	Colour		Spectral Distribution $E(\lambda)$	$E(\lambda) \cdot V(\lambda)_{day}$	$E(\lambda) \cdot V(\lambda)_{night}$
LED	Red	λ_{max} (nm)	636 ± 1	635 ± 1	630 ± 1
		FWHM (nm)	22 ± 1	22 ± 1	37 ± 1
	Yellow	λ_{max} (nm)	590 ± 1	590 ± 1	588 ± 1
		FWHM (nm)	16 ± 1	16 ± 1	21 ± 1
	Green	λ_{max} (nm)	496 ± 1	506 ± 1	500 ± 1
		FWHM (nm)	33 ± 1	40 ± 1	30 ± 1
	Blue	λ_{max} (nm)	460 ± 1	460 ± 1	460 ± 1
		FWHM (nm)	20 ± 1	28 ± 1	21 ± 1
Traditional	Red	λ_{max} (nm)	–	610 ± 1	593 ± 1
		FWHM (nm)	–	50 ± 1	36 ± 1
	Yellow	λ_{max} (nm)	–	574 ± 1	545 ± 1
		FWHM (nm)	–	88 ± 1	52 ± 1
	Green	λ_{max} (%)	–	544 ± 1	520 ± 1
		FWHM (nm)	–	62 ± 1	66 ± 1
	Blue	λ_{max} (%)	–	510 ± 1	480 ± 1
		FWHM (nm)	–	63 ± 1	46 ± 1

Note: $E(\lambda)$ = relative spectral distribution of luminance of the signal light, irrespective of vision sensitivity; $E(\lambda) \cdot V(\lambda)_{day}$ = relative spectral distribution of luminance of the signal light, weighted with vision sensitivity in daylight conditions; $E(\lambda) \cdot V(\lambda)_{night}$ = relative spectral distribution of luminance of the signal light, weighted with vision sensitivity in night-time conditions; λ_{max} = wavelength for which there is maximum energy of radiation; FWHM = full width at half maximum; resulting from approximation of the analysed distributions, accuracy is ±1 nm; – means that the distribution function nature does not allow to determine λ_{max} for a given colour and FWHM.

light distribution) range, weighted by vision sensitivity in daylight conditions ($E(\lambda) \cdot V(\lambda)_{day}$).

For traditional fluorescent signal lights:

- the shape of ($E(\lambda)$) radiation energy is "elongated". It is not a radiation peak, for which it is possible to determine full width at half maximum (FWHM). Only after taking into account (weighting) $E(\lambda)$ distributions by vision sensitivity in daylight or night-time condition distributions, the shape of spectral distributions is similar to that of Gaussian distribution. However, these distributions' full width at half maximum (FWHM) is significantly larger than distributions concerning LED sources, and it is between 50 ± 1 nm in red signal light weighted by vision sensitivity in daylight conditions ($E(\lambda) \cdot V(\lambda)_{day}$), to 88 ± 1 nm, in case of yellow signal light weighted by vision sensitivity in daylight conditions ($E(\lambda) \cdot V(\lambda)_{day}$), range.

- wavelength (λ_{max}), for which radiation maximum energy clearly shifts in the shorter wave direction with the use of vision sensitivity in night-time condition distribution. These shifts are: 17 ± 1 nm in case of red colour, 24 ± 1 nm in case of green colour, 29 ± 1 nm in case of yellow colour, and 30 ± 1 nm in case of blue colour.

On the basis of the analysis of the relative spectral distributions of luminance of the signal light emitted from LED type sources and traditional fluorescent sources, it may be assumed that the former will be more easily recognised, regardless of vision condition (day or night). However, this conclusion does not concern the case in which we look through colour filters that affect colour recognition significantly.

2.2.2.2.8 Coefficients for Quality Assessment of Colour Recognition

For the assessment of undisturbed colour vision while looking through optical protective filters, the following coefficients may be defined:

- signal lights transmittance factor for red, yellow, green, and blue colours (see formula 2.15),
- luminous transmittance weighted by spectral distribution of radiation of CIE standard illuminant D65 and spectral luminous efficiency function of the average human eye in daylight conditions and luminous transmittance weighted by spectral distribution of radiation of CIE standard illuminant A and spectral luminous efficiency function of the average human eye in night-time conditions (see formula 2.16),
- relative visual attenuation coefficient for signal light detection for red, yellow, green and blue colours in day and night vision conditions (see formula 2.17).

$$\tau_{\text{sign/dzień}} = \frac{\int_{380\,\text{nm}}^{780\,\text{nm}} \tau(\lambda)\cdot E(\lambda)\cdot V(\lambda)_{\text{dzień}}\, d\lambda}{\int_{380\,\text{nm}}^{780\,\text{nm}} E(\lambda)\cdot V(\lambda)_{\text{dzień}}\, d\lambda} \qquad \tau_{\text{sign/noc}} = \frac{\int_{380\,\text{nm}}^{780\,\text{nm}} \tau(\lambda)\cdot E(\lambda)\cdot V(\lambda)_{\text{noc}}\, d\lambda}{\int_{380\,\text{nm}}^{780\,\text{nm}} E(\lambda)\cdot V(\lambda)_{\text{noc}}\, d\lambda} \tag{2.15}$$

where:

$\tau_{\text{sign/day}}$ = transmittance factor of signal lights in daylight conditions,
$\tau_{\text{sign/night}}$ = transmittance factor of signal lights in night-time conditions,
$\tau(\lambda)$ = spectral transmittance,
$V(\lambda)_{\text{day}}$ = spectral luminous efficiency function of the average human eye in daylight conditions,
$V(\lambda)_{\text{night}}$ = spectral luminous efficiency function of the average human eye in night-time conditions,
$E(\lambda)$ = relative spectral distribution of luminance of the signal light, and
λ = wavelength.

$$\tau_{vD65/dzień} = \frac{\int_{380\,nm}^{780\,nm} \tau(\lambda) \cdot V(\lambda)_{dzień} \cdot S(\lambda)_{D65} d\lambda}{\int_{380\,nm}^{780\,nm} V(\lambda)_{dzień} \cdot S(\lambda)_{D65} d\lambda} \quad \tau_{vA/noc} = \frac{\int_{380\,nm}^{780\,nm} \tau(\lambda) \cdot V(\lambda)_{noc} \cdot S(\lambda)_{A} d\lambda}{\int_{380\,nm}^{780\,nm} V(\lambda)_{noc} \cdot S(\lambda)_{A} d\lambda} \tag{2.16}$$

where:

$\tau_{vD65/day}$ = luminous transmittance weighted by spectral distribution of radiation of standard illuminant D65 and spectral luminous efficiency function of the average human eye in daylight conditions,

$\tau_{vA/night}$ = luminous transmittance weighted spectral distribution of radiation of standard illuminant A and spectral luminous efficiency function of the average human eye in night-time conditions,

$\tau(\lambda)$ = spectral transmittance,

$V(\lambda)_{day}$ = spectral luminous efficiency function of the average human eye in daylight conditions,

$V(\lambda)_{night}$ = spectral luminous efficiency function of the average human eye in night-time conditions,

$S(\lambda)_{D65}$ = spectral distribution of radiation of CIE standard illuminant D65,

$S(\lambda)_A$ = spectral distribution of radiation of CIE standard illuminant A, and

λ = wavelength.

$$Q_{sign/dzień} = \frac{\tau_{sign/dzień}}{\tau_{vD65/dzień}} \quad Q_{sign/dzień} = \frac{\tau_{sign/noc}}{\tau_{vA/noc}} \tag{2.17}$$

where:

$Q_{sign/day}$ = relative visual attenuation coefficient for signal light detection in daylight conditions,

$Q_{sign/night}$ = relative visual attenuation coefficient for signal light detection in night-time conditions,

$\tau_{sign/day}$ = transmittance factor of signal lights in daylight conditions,

$\tau_{sign/night}$ = transmittance factor of signal lights in night-time conditions,

$\tau_{vD65/day}$ = luminous transmittance weighted by spectral distribution of radiation of standard illuminant D65 and spectral luminous efficiency function of the average human eye in daylight conditions, and

$\tau_{vA/night}$ = luminous transmittance weighted spectral distribution of radiation of standard illuminant A and spectral luminous efficiency function of the average human eye in night-time conditions.

In order to analyse the possible colour vision abilities while looking through colour filters, the value of relative visual attenuation coefficient for signal light detection is of a fundamental meaning. This coefficient (see formula 2.17) is determined by a relation of transmittance factor of a given signal light in determined vision conditions (day or night) value to the value of luminous transmittance, which is determined with regard to spectral distribution of radiation of illuminant and spectral efficiency function of the eye, adjusted to given vision conditions (day or night). If a signal light is observed during the day, luminous transmittance is determined with regard to D65 illuminant and spectral luminous efficiency function of the eye in daylight conditions. If a signal light is observed during the night, luminous transmittance is determined with regard to A illuminant and spectral luminous efficiency function of the eye in night-time conditions. Luminous transmittance that is taken into account while calculating relative visual attenuation coefficient for signal light detection depending on day or night vision conditions is calculated in relation to conditions in which an eye was adapted.

If the value of relative visual attenuation coefficient for signal lights is less than 1, it means that a light beam that takes into account a certain colour signal light spectral distribution, which transmits through filter, is smaller than the light beam transmitted through filter, for which the spectral distribution of a given signal light has not been taken into account. If the value of relative visual attenuation coefficient for signal lights is more than 1, it means the opposite situation. Paradoxically, a relative visual attenuation coefficient for signal light should be determined as "reinforcement coefficient" for signal lights in such a case. In case of analysing relative visual attenuation coefficient for signal light changes (resulting from, e.g., using different spectral power distributions of illuminant in calculations or signal lights only), the positive change (+) means that signal lights will be possibly more recognisable. The negative change (–) means that signal light vision will be potentially worse.

2.2.2.2.9 Transmittance Factors in Melatonin Secretion Assessment

Secreted by the pineal gland into the blood stream, melatonin informs the body about night-time processes, such as sleep or lowering body temperature and slowing heart rate. These kind of rhythms in the human body including the sleep and awareness cycle; concentration, efficiency and mood changes during the day; reactions to seasonal changes are all dependent on light conditions. One method of melatonin suppression is exposure to daylight of the right colour and intensity (425–560 nm range), and especially to blue light of about 465 nm wavelength. This electromagnetic radiation range is characterised by maximum efficiency of melatonin suppression process [Aube 2013; Brainard 2005; Thapan 2001].

In the formulae for calculating weighted transmittance factors, a spectral distribution function is taken into account. This function may reflect a certain biological effect which refers to vision efficiency. In the melatonin suppression process effect, the 425–560 nm range of wavelengths is taken into account. Spectral distribution function, which will be used to determine optical radiation transmittance factor with regard to this effect, should correspond to the melatonin suppression process. In order to find a coefficient which will determine the melatonin suppression process efficiency, three different spectral distributions of melatonin suppression process efficiency have been taken into account – według Brainarda et al. [2001], Aube et al. [2013], and Thapan et al. [2001]. The described spectral distributions – prepared on the basis of the mentioned above citations – are presented in Figure 2.43.

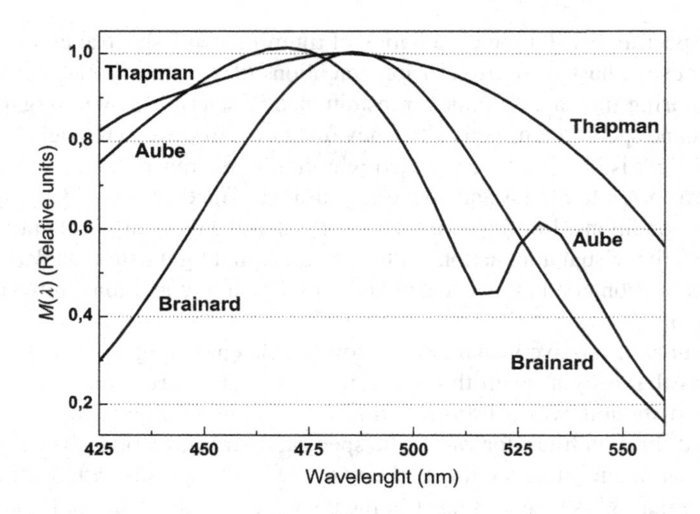

FIGURE 2.43 Relative spectral distribution of melatonin suppression process efficiency ($M(\lambda)$).

Optical radiation transmittance factor for 425–560 nm wavelength range, which takes into account melatonin suppression process efficiency spectral distribution, is defined by the following formula [Owczarek et al. 2017]:

$$\tau_M = \frac{\int_{425\,\mathrm{nm}}^{560\,\mathrm{nm}} \tau(\lambda) \cdot M(\lambda) d\lambda}{\int_{425\,\mathrm{nm}}^{560\,\mathrm{nm}} M(\lambda) d\lambda} \tag{2.18}$$

where:

τ_M = transmittance factor taking into account spectral distribution of efficiency of melatonin suppression process,

$\tau(\lambda)$ = spectral transmittance expressed in percents,

$M(\lambda)$ = melatonin suppression efficiency relative spectral distribution (according to Brainard et al., Aube et al., lub Thapana et al.), and

λ = wavelength.

Transmittance factor, defined by the above presented formula, determines how the optical radiation of 425–560 nm range, falling on the filter surface, is alleviated in relation to melatonin suppression process efficiency. In order to take the average value of the above-mentioned distribution into account, the following formula may be used:

$$\tau_{Ms} = \frac{\tau_{MB} + \tau_{MA} + \tau_{MT}}{3} \tag{2.19}$$

where:

τ_{Ms} = mean value of the transmittance factor that takes into account melatonin suppression process efficiency spectral distribution,

τ_{MB} = transmittance factor of visible radiation of 425–560 nm wavelength range that takes into account melatonin suppression process efficiency spectral distribution, according to Brainard et al.,

τ_{MA} = transmittance factor of visible radiation of 425–560 nm wavelength range that takes into account melatonin suppression process efficiency spectral distribution, according to Aube et al., and

τ_{MT} = transmittance factor of visible radiation of 425–560 nm wavelength range that takes into account melatonin suppression process efficiency spectral distribution, according to Thapan et al.

The ratio of difference between a total amount of visible radiation transmitting through the filter (τ_v) and efficient amount of visible radiation of 425–560 nm range (τ_{Ms}) to total amount of visible radiation (τ_v) transmitting through the optical protective filter has been determined as *P*, which is expressed by the following formula:

$$P = \frac{\tau_v - \tau_{Ms}}{\tau_v} \times 100\% \tag{2.20}$$

where:

P = a coefficient determining a ratio of difference between a total amount of visible radiation transmitting through the filter and efficient amount of visible radiation of 425–560 nm range to total amount of visible radiation transmitting through visible radiation filter,

τ_v = transmittance factor determining the total amount of visible radiation transmitted through a given filter, and

τ_{Ms} = mean value of transmittance factor that takes into account melatonin suppression process efficiency spectral factor.

The factors defined above can be used for assessment of possible influence of filter colour on melatonin secretion.

2.2.3 Eye Protectors for Individuals with Vision Dysfunction

2.2.3.1 Vision Dysfunction

The inability to see properly can lead to exclusion from social and, above all, professional life. According to a report published in 2016 by the Institute for Health Protection [Mariotti 2012; Raciborski 2016], the most common causes of vision disorders in the world include refractive errors (42%), cataracts (33%), and glaucoma (2%). The dominating causes behind blindness are cataracts (51%), followed by glaucoma (8%) and macular degeneration, called AMD – *age-related macular degeneration* (5%). Eye disorders escalate with age and are an inseparable result of an ageing body [AAO 2013]. Both in Poland and across the globe, the number of people with eye disorders (cataracts, glaucoma, AMD, retinal complications in diabetes, embolisms and thrombi in retinal vessels) is growing rapidly. For example, AMD primarily affects people who are aged above 50, and it is a main cause of eye disability in the industrial world and the third leading cause worldwide [Wong 2014]. According to published data [Pennington 2016], about 11 million people are affected by AMD in the United States alone. Because ageing is the biggest risk factor, it is expected that the number of people affected by AMD in the USA will grow to 22 million by 2050.

According to a healthy ageing report [Fundacja na Rzecz Zdrowego Starzenia się 2015], this issue affects about 1.5 million people in Poland. However, it is worrying that the eye disability issue is affecting increasingly younger people, for instance, cataract surgeries are also performed on people aged less than 50, and even on teenagers. The causes for this state of things should be looked for in environmental impacts, such as spending more and more time in front of monitors and screens of electronic devices (intense visual work), staying long in air-conditioned rooms (exposure to biological factors), poor diet, and many others. It should be noted that diseases pertinent to eyes is a global issue and it affects a large number of people, also those of working age. Benefits provided by Social Insurance Institution (ZUS) related to inability to work within the group: eye diseases and adnexa oculi increased by 18.3% (1805.3 thousand) in Poland in 2016 when compared to 2012 (1474.8 thousand) [ZUS 2017]. According to the *European Health Interview Survey* (EHIS) studies conducted by Statistics Poland (GUS) in 2014 (latest data published by the Healthcare Institute in 2016), more than 52% of the Polish population aged 15 and above used spectacles or contacts in order to correct errors of refraction. Within the age group of 50–59 years, this number reached nearly 76% and in the next three age groups above 80% [The Healthcare Institute 2016]. Current data published by the WHO confirm that errors of refraction are the most common type of occurring disorder, and furthermore, they are not age related [Hashemi 2018].

Eye disorders, depending on their type and stage, may affect quality and comfort of the performed work or, in extreme cases, be an impediment in occupying certain professions. Eye disorders that demand correction can significantly limit access to perform specific professional activities based on precise visual tasks where full range of sharp vision is necessary. Colour vision deficiency excludes people especially from occupying professions in the field of transport or medicine [Information materials from szkla.com 2019]. It should be clearly underlined that people with eye disorders must be treated in specific ways in their working environment for work safety reasons.

Depending on the eye disorder type, adequate eye protectors shall be used, especially in case of special eye protectors in the form of spectacles and protective googles. In the qualification process of persons with eye disorders for workstations where eye protectors are necessary, special attention should be paid where the used personal protective equipment should ensure both protection and individual demand for level of vision and comfort in visual work.

2.2.3.2 Spectacles and Goggles with Corrective Effect

On the market of eye protectors exist special eye protectors for professionally active people with errors of refraction. Under the currently binding law, all personal protective equipment, including this dedicated for users that need vision correction, should be admitted to use on the basis of confirming their compliance with essential requirements consisted in 425/2016 regulation on personal protective equipment [Regulation of the European Parliament and the EU Council 2016]. There are several solutions for eye protectors for persons with errors of refraction, by using: eye protectors adjusted for wearing them together with corrective spectacles or contacts; special eye protectors equipped with corrective inserts; and corrective-protective dual-function spectacles, and goggles. It is very important that the correction used in eye protectors should be individually matched to the user's needs, which makes the personal protective equipment a dedicated item – from then on it can be only used by this individual user. Within eye protectors dedicated for persons with errors of refraction (protective spectacles and goggles), those that allow for installing corrective lens for an individual user can be distinguished [Hayne Polska LLC informational materials]. Examples of protective spectacles with a special corrective insert, designed especially for persons with errors of refraction, are shown in the Figure 2.44.

These corrective (prescription) spectacles and corrective inserts distribution, due to the necessity of matching the correction individually, is done through specialised companies or at an optician's office.

Another design solution is to use protective spectacles and goggles together with corrective spectacles. In this case, eye protectors are worn over corrective spectacles [3M informational materials]. Protective spectacles that are worn over the corrective ones are designed with regard to such overmeasures as to fit most typical designs of prescription spectacles. In this type of protective spectacles, it is

FIGURE 2.44 Protective spectacles that allow for installing a corrective insert.

(Continued)

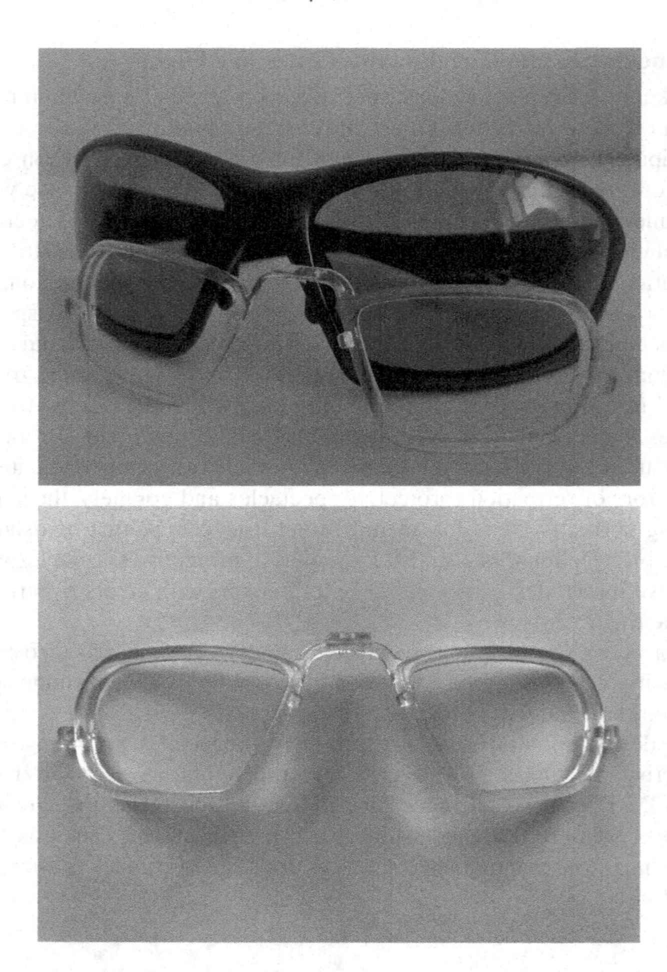

FIGURE 2.44 (Continued) Protective spectacles that allow for installing a corrective insert.

especially important that the housing design shall allow their fitting to the face of a user that already wears prescription spectacles. Comfortable use is additionally provided by 4-stages length adjustment for soft, low-profile temples, where interaction with corrective spectacles is minimal. In addition, lenses' adjustable pitch also allows for an easy fit and a high level of comfort of use. A precise fitting guarantees compatibility with many types of prescription spectacles. For persons with errors of refraction, there are also dual-function spectacles available; they are both protective and corrective and the lenses in them are installed permanently. Examples of this solution are shown in Figure 2.45.

Today, there are companies on the market that offer dual-function, protective-corrective spectacles, in which a corrective lens is an integral part of protective spectacles lenses. Depending on the solution, these are progressive lenses or bifocal

FIGURE 2.45 Examples of protective spectacles with corrective lenses.

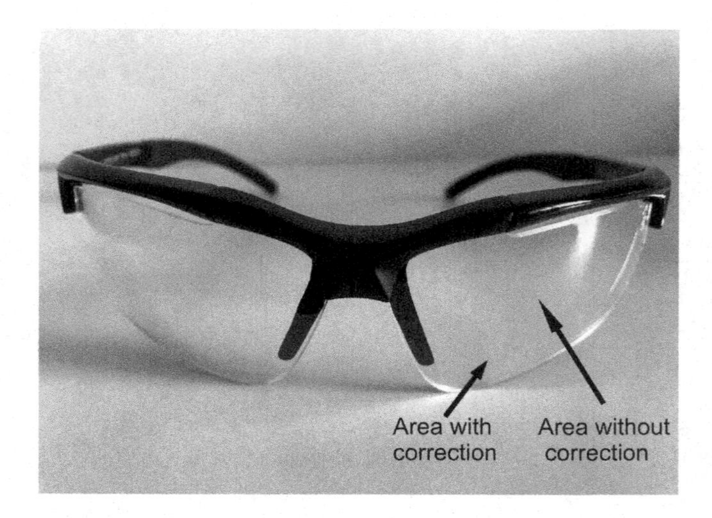

FIGURE 2.46 Protective spectacles with corrective additive lens.

lenses with additive correction, which means that the upper part of protective spectacles can be an area with distance correction or zero correction, while the lower or middle part is an area with an additive [informational materials from Nezo company]. The type of spectacles with an additive is shown in Figure 2.46.

In this case (presented in Figure 2.46), these are bifocal lenses, where the upper part is an area without correction, and the lower part prescription glasses, with the optical power of +1,5; +2; +2,5 dioptre, allows to magnify an image, and they are useful in reading or precise works. Specially selected frame shapes with full adjustment, both in horizontal and vertical levels, allow a very good fit to the shape of the head. In these types of designs, spaces between spectacles and the face are minimised due to the use of many stages adjustment, which in practice limits the risk of dangerous particles reaching the eye. Lenses made of polycarbonate in the first optical classes are designed for a permanent wear, and the designer guarantees lack of dizziness or vision deformation. Such frames, equipped with flat temples – useful when wearing hearing protectors – and a silicone nose pad, provide user's comfort. Protective-corrective spectacles can be equipped with any lens type: unifocal, bifocal or progressive [Informational materials from Safetyline company]. Some companies manufacturing these protective spectacles offer product acquisition combined with eye examination carried out in the employer's company.

2.2.3.3 Special Optical Protective Filters for Individuals Using Intraocular Lenses (IOLs)

Nowadays, many people have artificial *interocular lenses* (IOL) implemented. Some people also underwent surgical correction of error of refraction and have undergone cataract removal surgery or lost their lens because of mechanical injuries. It should be emphasised that the number of people with IOLs is growing, and the average age of patients undergoing eye disorder correction surgeries, such as cataract removal, is decreasing. This creates a large group of people of working age for whom there shall be ensured adequate conditions of visual work.

Replacing the natural lens in an eye with an artificial one may cause a change in the level at which the optical filter shall block harmful optical radiation before it reaches the retina. The colour of the intraocular lens may also cause a change in colour recognition ability, which is a very important function while performing visual work (i.e., signal lights recognition, technological processes observation, etc.).

While an intraocular lens blocks ultraviolet radiation (UV), which is dangerous for intraocular structures (e.g., retina), practically in total – similar to conditions provided by human crystalline lens –its blocking of luminous transmittance in full range of its spectral characteristic results in IOL lens significantly differing from a natural lens [Owczarek 2012, 2016]. From a technical viewpoint, modification of light transmitted in IOLs consists in introducing into the materials, of which the lenses are made, colours (chromophores) limiting luminous transmittance of a determined spectral range. A yellow chromophore, that is a blue colour filter, is introduced most often. This modification, similar to the case of filters or lenses used in corrective spectacles, results in changes at the level of optical radiation transmittance.

Decreasing the level of luminous transmittance in IOL lens as a result if its usage involves many factors, the analysis of which is a separate, extensive research issue. One of the most important causes of this fact is the common occurrence of the glistening phenomenon. This phenomenon is common in acrylic and hydrophobic IOLs. The relation of glistening to refractive IOL value, and thus to its width, and also to smaller cartridge diameter during IOL inoculation may indicate that a potential damage of IOL has important meaning in glistening. Lower density of glistening in patients undergoing intensive anti-inflammatory treatment during diabetes and those with uvea inflammation may indicate that the phenomenon is related to rupture of the ventriculovascular barrier [Godlewska 2016; Jurowski 2012]. A microphotograph of the explanted, examined hydrophilic lens which illustrates glistening phenomenon is shown in Figure 2.47. In the figure, numerous microvacuoles, concentrated in central (optical) lens part, can be seen.

In order to determine the spectral characteristics of filter + IOL system with eliminated decrease of transmittance level associated with the method of spectrophotometric measurement, a mathematical method of determining spectral characteristic of the filter + IOL system was proposed. It consists of separate determination of spectral characteristics for ($\tau(\lambda)_F$) filter and ($\tau(\lambda)_{IOL}$) IOL lens, and next, calculation of transmittance factor of ($\tau(\lambda)_{F+IOL}$) filter + IOL system. Figure 2.48 shows the mathematical method of determining spectral characteristic of the filter + IOL system.

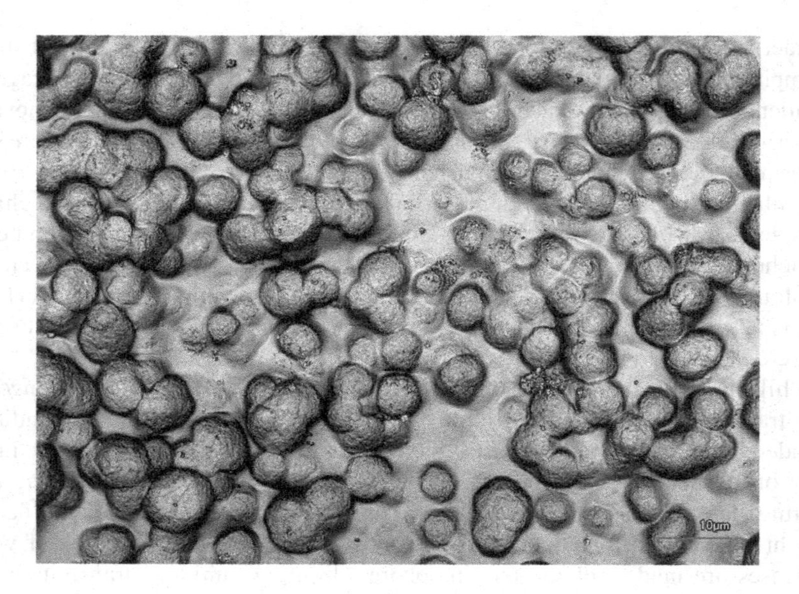

FIGURE 2.47 Microvacuoles in a central part of the explanted hydrophilic IOL.

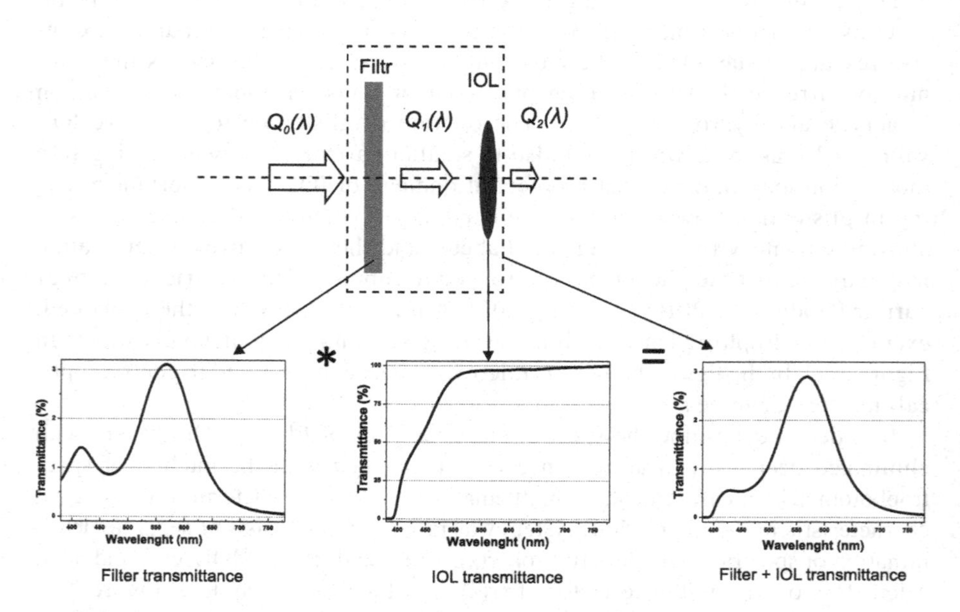

FIGURE 2.48 A scheme presenting the mathematical method of determining spectral characteristic of the filter + IOL system. $Q_0(\lambda)$ – light beam falling on the filter, $Q_1(\lambda)$ – light beam falling on IOL after absorbing by the filter, $Q_2(\lambda)$ – light beam passing through the filter + IOL system.

In case of filter + IOL system, the spectral luminous factors for IOL and filter are expressed by the 2.21 and 2.22 formulae.

$$\tau(\lambda)_{IOL} = \frac{Q_1(\lambda)}{Q_0(\lambda)}, \tag{2.21}$$

$$\tau(\lambda)_{F} = \frac{Q_2(\lambda)}{Q_1(\lambda)}, \tag{2.22}$$

where:

$Q_0(\lambda)$ = light beam falling on the filter,
$Q_1(\lambda)$ = light beam falling on IOL after absorbing by the filter,
$Q_2(\lambda)$ = light beam passing through filter + IOL system, and
λ = wavelength.

From the 2.21. and 2.22 formulae, we obtain a formula for spectral transmittance of the filter + IOL system, which is expressed by formula 2.23.

$$\tau(\lambda)_{F+IOL} = \frac{Q_2(\lambda)}{Q_0(\lambda)} = \frac{\tau(\lambda)_F \cdot Q_1(\lambda)}{Q_0(\lambda)} = \tau(\lambda)_F \cdot \tau(\lambda)_{IOL} \tag{2.23}$$

In the 2.23 formula, it can be seen that spectral transmittance factor of ($\tau(\lambda)_{F+IOL}$) filter + IOL system is a product of spectral transmittance factors ($\tau(\lambda)_{IOL}$) IOL and ($\tau(\lambda)_F$) filter. Thus, the 2.23 formula allows us to determine the IOL lens spectral characteristic shape influence on the change of transmittance through the filter + IOL system in relation to the transmittance of the filter itself for any IOL lens; the spectral characteristic is determined by the lens designer.

2.2.4 New Trends in Designing Eye and Face Protectors

The development of new technologies and innovative materials imposes directions for the development of modern and future protective eye and face equipment. New trends in the design of protective eye and face equipment include the following four key areas:

- materials engineering,
- electronics and information technology,
- ergonomics, and
- design.

Modern protective eye and face equipment are made of lighter and more durable materials. Their features, such as resistance to fogging, significantly increase the protection and use value of designed spectacles, goggles or face shields. The weight reduction, while maintaining the required mechanical strength parameters, significantly increases the comfort of use. The use of thin-film technology to produce optical protective filters allows design of filters with characteristics tailored to the conditions of the hazard, workplace lighting and individual requirements of users. The achievements of material engineering in the production of optical filters, combined with the possibilities of electronics, enable the design and production of optical protective filters that automatically adjust the darkening to the intensity of harmful optical radiation. The equipment of eye and face protectors with cameras in connection with the use of information technology makes them become elements of the space, referred to as the Internet of Things (IoT). In this field, various types of physical objects are equipped with sensors and devices initiating specific actions (actuators). In this approach, it is the interrelated physical objects that set new areas of Internet functioning. They include data collection, remote monitoring, decision support algorithms, optimisation of various processes, etc. [Dobbs 2015].

The data that are collected by electronic systems integrated with eye and face protectors are mainly image data. The history of displaying an image directly in front of the user's eyes dates back to 1968. It was then that the first *A head-mounted three dimensional display* was developed [Sutherland 1968]. In the 1990s, the term augmented reality (AR) entered into general circulation. The author of this concept is Tom Caudell, who in 1992 created a system for Boeing's employees to facilitate the assembly of wires using this technology [Caudell 1992]. Augmented reality systems allow for the introduction of additional information in the user's field of view, graphically applied to the device with which the reality is observed. Nowadays such systems are widely available. One of the most famous projects in this area is a project developed by Google. The augmented reality Google Glass is to have the ultimate smartphone function and be operated with voice commands.

As a natural consequence, augmented reality systems can be used in the construction of any type of eye and face protector. When analysing technical solutions used in Google Glass [Google Materials] or other glasses of this type, it is easy to see that their construction is modular. This simply means that the camera and projector can be installed on any eyepiece frame, including spectacles, goggles or face shields. The technology is therefore no longer an obstacle to the use of augmented reality in protective eye and face equipment. Eye and face protectors designers with integrated augmented reality modules focus on selecting the range and form of the information displayed in AR systems so that this information provides actual help during the job. Research in this area was also conducted by the authors – the augmented reality system was installed in the welding safety helmet.

With regard to the designed graphical interface, utility tests were conducted to check, among other things, the welders' reactions to the appearance of the information displayed in the augmented reality system. A check of welders' reactions to the appearance of the information displayed in the augmented reality system was conducted in simulation tests in which the participants indicated the need to configure the graphical interface themselves and to limit the amount of information and warnings displayed by the AR system to those considered necessary. Overloading of information that occurs during the course of the work is disadvantageous because it distracts or misleads the user (Owczarek and Gralewicz 2014).

The use of innovative materials in connection with the results of the latest anthropometric research allows design of ergonomic constructions, which, by means of easy-to-use adjustment systems, can be used by a wide range of recipients. The design is also not without significance. Protective eye and face equipment of an interesting design is much more likely to be used.

REFERENCES

3M Poland Sp. z o.o.'s information materials. www.3mpolska.pl/3M/pl_PL/firma-pl/all-3m-products/~/3M-Seria-2800-2802-Okulary-ochronne-nak%C5%82adane-na-okulary-korekcyjne/?N=5002385+8709322+8711017+8711405+8720539+8720549+8727587+8738262+3294271826&rt=rud. (accessed September 20, 2019).

AAO [American Academy of Ophthalmology]. 2013. Data on US eye disease statistics. www.aao.org/eye-disease-statistics. (accessed September 20, 2019).

ANSI/ISEA [American National Standards Institute (ANSI)/International Safety Equipment Association (ISEA)]. 2014. Industrial head protection ANSI/ISEA Z89.1-2014. Washington, DC.

AS/NZS [Standards Australia, and Standards New Zealand]. 1997. Occupational protective helmets AS/NZS 1801:1997. Homebush, Australia and Wellington, New Zealand: AS/NZS.

Aube, M., J. Roby, and M. Kocifaj. 2013. Evaluating potential spectra impacts of various artificial lights on melatonin suppression, photosynthesis, and star visibility. *PLoS One* 8(7):1–15.

Baszczyński, K. 2014a. Przemysłowe hełmy ochronne a zabezpieczenie głowy przed uderzeniami bocznymi [Industrial safety helmets and protection of the head against side impacts]. *Bezpieczeństwo Pracy – Nauka i Praktyka* 5(370):10–13.

Baszczyński, K. 2014b. The effect of temperature on the capability of industrial protective helmets to absorb impact energy. *Eng. Fail. Anal.* 46:1–8.

Baszczyński, K. 2018. Analiza zagrożeń uszkodzeń głowy pracownika podczas powstrzymywania spadania z wysokości [Risk analysis of employee head injury during a fall arrest]. *Bezpieczeństwo Pracy – Nauka i Praktyka* 2(557):20–23.

Brainard, G. C., and J. P. Hanifin. 2005. Photons, clocks, and consciousness. *J. Biol. Rhythms.* 20(4):314–325.

Brainard, G. C., J. P. Hanifin, and J. M. Greeson et al. 2001. Action spectrum for melatonin regulation in humans: Evidence for a novel circadian photoreceptors. *J. Neurosci.* 21(16):6405–6412.

Caudell, T. P., and D. W. Mizell. 1992. Augmented reality: an application of heads-up display technology to manual manufacturing processes. *Proceedings of the Twenty-Fifth Hawaii International Conference on System Sciences.* Vol. 2. Kauai, HI. 659–669. doi:10.1109/HICSS.1992.183317.

CEN [European Committee for Standardization]. 1995. Personal eye-protection – Vocabulary. EN 165:1995. Brussels, Belgium: CEN.

CEN [European Committee for Standardization]. 2001. Personal eye-protection – Specifications. EN 166: 2001. Brussels, Belgium: CEN.

CEN [European Committee for Standardization]. 2002. Personal eye-protection – Ultraviolet filters – Transmittance requirements and recommended use. EN 170:2002. Brussels, Belgium: CEN.

CEN [European Committee for Standardization]. 2009. Personal eye-protection – Automatic welding filters (Specification for welding filters with switchable luminous transmittance and welding filters with dual luminous transmittance). EN 379:2003+A1:2009. Brussels, Belgium: CEN.

CEN [European Committee for Standardization]. 2012a. Industrial safety helmets. EN 397:2012+A1:2012. Brussels, Belgium: CEN.

CEN [European Committee for Standardization]. 2012b. High performance industrial helmets EN 14052:2012+A1:2012. Brussels, Belgium: CEN.
CEN [European Committee for Standardization]. 2012c. Industrial bump caps EN 812:2012. Brussels, Belgium: CEN.
CIE [Commission Internationale de l'Eclairage]. 1926. Photopic luminosity function. *CIE Proceedings*. Cambridge: Cambridge University Press.
CIE [Commission Internationale de l'Eclairage]. 1951. Scotopic luminosity curve. Paris: CIE Proceedings. Bureau Central de la CIE. Geneva, Switzerland: CIE.
CIE [Commission Internationale de l'Eclairage]. 1999. Technical Report. A Method for Assessing the Quality of Daylight Simulators for Colorimetry. 51.2-1999 (including Supplement 1-1999). Paris, France: Bureau central de la CIE.
Crawford, B. H. 1949. The scotopic visibility function. *Proc. Phys. Soc.* B62:321–334.
CSA [Canadian Standards Association]. 2015. Eye and Face Protectors. CSA Z94.3-15. Standards Council of Canada (SCC).
Dobbs, R., J. Manyika, and J. Woetzel. 2015. *No Ordinary Disruption: The Four Global Forces Breaking All the Trends*. New York: Public Affairs.
Directive 2006/25/EC of the European Parliament and of the Council of 5 April 2006 on the minimum health and safety requirements regarding the exposure of workers to risks arising from physical agents (artificial optical radiation) (19th individual Directive within the meaning of Article 16 (1) of Directive 89/391/EEC). Official Journal of the European Union L 114/38.
Forero Rueda, M. A., L. Cui, and M. D. Gilchrist. 2009. Optimisation of energy absorbing liner for equestrian helmets. Part I: Layered foam liner. *Mater. Des.* 30:3405–3413.
Fundacja na Rzecz Zdrowego Starzenia się. 2015. Raport na rzecz zdrowego starzenia się. Okulistyka i choroby siatkówki w aspekcie zdrowego i aktywnego starzenia się. https://fpbb.pl/user_upload/static/file/Materia%C5%82y%202015/Raport_okulistyka_i_choroby_siatkowki.pdf (accessed September 20, 2019).
Gilchrist, A., and N. J. Mills. 1989. Improved side, front and back impact protection for industrial helmets. School of Metallurgy and Materials. University of Birmingham. Health & Safety Executive contract research report no. 13/1989.
GNU Image Manipulation Program (GIMP). 2012. Technical materials 2.10.12 Version. www.gimp.org/. (accessed September 20, 2019).
Godlewska, A., G. Owczarek, and P. Jurowski. Zjawisko "glistening" w sztucznych akrylowych hydrofobowych soczewkach wewnątrzgałkowych – jak na częstość jego występowania i nasilenie wpływają czynniki okołooperacyjne i choroby mu towarzyszące. *Klinika Oczna* 3:191–196.
Google Materials, Technical materials. www.google.com/glass/start. (accessed September 20, 2019).
Hashemi, H., A. Fotouhi, A. Yekta, R. Pakzad, H. Ostadimoghaddam, and M. Khabazkhoob. 2018. Global and regional estimates of prevalence of refractive errors: Systematic review and meta-analysis. *J. Curr. Ophthalmol.* 30(1):3–22.
Hayne Polska Sp. z o.o.'s information materials. www.hayne.pl/pl/okulary-ochronne/2964-okulary-ochronne-z-wkladka-korekcyjna-h1001100-h1001100.html. (accessed September 20, 2019).
Hui, S. K., and T. X. Yu. 2002. Modelling of the effectiveness of bicycle helmets under impact. *Int. J. Mech. Sci.* 44(6):1081.
Hulme, A. J., and N. J. Mills. 1996. The performance of industrial helmets under impact. An assessment of the British standard BS 5240 PT. 1, 1987. School of Metallurgy and Materials. University of Birmingham. Health & Safety Executive contract research report no. 91/1996.
ISO [International Organization for Standardization]. 1977. Industrial safety helmets. ISO 3873 1st Edition 1977. Geneva, Switzerland: ISO.

ISO [International Organization for Standardization]. 1999. CIE standard illuminants for colorimetry (Standards No. ISO/CIE 10526: 1999). Geneva, Switzerland: CIE.

ISO [International Organization for Standardization]. 2001. Ophthalmic optics. Spectacle lenses. Vocabulary. PN-EN ISO 13666:2001. Geneva, Switzerland: ISO.

ISO [International Organization for Standardization]. 2014. Personal protective equipment. Test methods for sunglasses and related eyewear. PN-EN ISO 12311: 2014. Geneva, Switzerland: ISO.

Jurowski, P., K. Kaczorowska-Rusin, and G. Owczarek. 2012. Współczesny stan wiedzy na temat zjawiska "glistening" w sztucznych soczewkach wewnątrzgałkowych. *Klinika Oczna* 4:317.

Korycki, R. 2002. The damping of off-central impact for selected industrial safety helmets used in Poland. *Int. J. Occup. Saf. Ergo.* 1(8):51–70.

Lens, A., S. Coyne Nemeth, and J. K. Ledford. 2010. *Anatomia i fizjologia narządu wzroku.* Polish edition M. Misiuk-Hojło. Wrocław: Górnicki Wydawnictwo Medyczne.

Mariotti, S. P., and D. Pascolini. 2012. Global estimates of visual impairment 2010. *Br. J. Ophthalmol.* 96(5):614–618.

Mills, N. J., and A. Gilchrist. 1990. Proposals for side-impact and retention tests for industrial helmets. School of Metallurgy and Materials. Birmingham, UK: University of Birmingham.

Morgan, E. 2018. Eyeglass frame materials: Metal, plastic and unusual. www.allaboutvision.com/eyeglasses/eyeglass_frame_materials.htm. (accessed September 20, 2019).

Nezo's information materials. www.nezo.pl/files/media/zdjecia/bxreaderbaner.jpg. (accessed September 20, 2019).

Owczarek, G. 2001. Sprawozdanie z projektu 03.8.18. Opracowanie wytycznych dla optymalizacji konstrukcji automatycznych filtrów spawalniczych, uwzględniających rzeczywiste warunki użytkowania. Warszawa: Centralny Instytut Ochrony Pracy – Państwowy Instytut Badawczy.

Owczarek, G. 2013. Report on statutory task CIOP-PIB. Badania nad zastosowaniem systemów rzeczywistości wzbogaconej dla osób z dysfunkcja narządu wzroku (Research on the application of augmented reality systems for people with visual impairment). Warszawa: Centralny Instytut Ochrony Pracy – Państwowy Instytut Badawczy.

Owczarek, G., G. Gralewicz, N. Skuza, and P. Jurowski. 2016. Light transmission through intraocular lenses with or without yellow chromophore (blue light filter) and its potential influence on functional vision at everyday environmental conditions. *Int. J. Occup. Saf. Ergo.* 1(22):66–70.

Owczarek, G., and G. Gralewicz. 2014. Sposób prezentacji graficznej informacji wyświetlanych w modelu rzeczywistości wzbogaconej zintegrowanym z przyłbicą spawalniczą. *Works of the Institute of Electrical Engineering* 264:15–29.

Owczarek, G., and P. Jurowski. 2012. Zmiany transmisji promieniowania optycznego przez soczewki wewnątrzgałkowe eksplantowane z powodu zjawiska glisteningu. *Prace Instytutu Elektotechniki* (*Works of the Institute of Electrical Engineering*) 255:201–211.

Owczarek, G., G. Gralewicz, A. Wolska, N. Skuza, and P. Jurowski. 2017. Potencjalny wpływ barwy filtrów w okularach chroniących przed olśnieniem słonecznym na wydzielanie melatoniny. *Med. Pr.* 68(5):629–637.

Pennington, K. L., and M. M. DeAngelis. 2016. Epidemiology of age-related macular degeneration (AMD): Associations with cardiovascular disease phenotypes and lipid factors. *Eye Vision* 3(34):1–20.

Pinnoji, P. K., P. Mahajan, N. Bourdet, C. Deck, and R. Willinger. 2010. Impact dynamics of metal foam shells for motorcycle helmets: Experiments & numerical modeling. *Int. J. Impact. Eng.* 37(3):274.

Pościk, A., J. Kubrak, and L. Włodarski. 2006. Barwniki fotochromowe, interferencja światła i automatyczne filtry spawalnicze. *Prace Instytutu Elektotechniki* (*Works of the Institute of Electrical Engineering*) 228:155–166.

Raciborski, F., A. Kłak, E. Gawińska, and M. Figurska. 2016. Institute for Health Protection. Choroby oczu – problem zdrowotny, społeczny oraz wyzwanie cywilizacyjne w obliczu starzenia się populacji, Raport Instytutu Ochrony Zdrowia. Warszawa: Instytut Ochrony Zdrowia. https://spartanska.pl/wp-content/uploads/ raport_choroby_oczu.pdf (accessed September 20, 2019).

Reviews & Buying Guide – 26 Best Welding Helmets. 2019. https://bestweldinghelmet.review/. (accessed September 20, 2019).

Regulation of the European Parliament and the EU Council. 2016. Regulation (EU) 2016/425 of the European Parliament and of the Council of 9 March 2016 on personal protective equipment and repealing Council Directive 89/686/EEC. https://eur-lex.europa.eu/eli/reg/2016/425/oj. (accessed September 20, 2019).

Safetyline's information materials. http://safetyline.pl/oferta/okulary-ochronne-korekcyjne/. (accessed September 20, 2019).

Sharpe, L. T., A. Stockman, H. Jägle, and J. Nathans, 1999. Opsin genes, cone photopigments, color vision and color blindness. In *Color Vision: From Genes to Perception*, ed. K. R. Gegenfurtner, and L. T. Sharpe, 1–51. Cambridge: Cambridge University Press.

Shuaeib, F. M., A. M. S. Hamouda, M. M. Hamdan, R. S. Radin Umar, and M. S. J. Hashmi. 2002. Motorcycle helmet. Part II. Materials and design issues. *J. Mater. Process Technol.* 123:422–431.

Shuaeib, F. M., A. M. S. Hamouda, S. V. Wong, R. S. Radin Umar, and M. M. H. Megat Ahmed. 2007. A new motorcycle helmet liner material: The finite element simulation and design of experiment optimization. *Mater. Des.* 28:182–195.

Sutherland, I. E. 1968. A head-mounted three dimensional display. *Proceedings of the AFIPS '68. Fall Joint Computer Conference, Part I*, December 9–11. ACM. 757–764. Washington, DC: Thompson Books.

szkla.com's information materials. 2019. Poradnik Optometrysty (Optometrist's Guide). Available at www.szkla.com/poradnik_kondycja-oczu-a-praca-w-zawodzie.html. (accessed September 20, 2019).

Thapan, K., J. Arendt, and D. Skene. 2001. An action spectrum for melatonin suppression: Evidence for a novel non-rod, non-cone photoreceptor system in humans. *J. Physiol.* 535:261–267.

Users Guide. 2006. PPG Industries, Inc. Trivex PGE Industries, The Tri-Performance Lans Materials. www.premiumoptics.com/Downloads/Trivex_Users_Guide.pdf. (accessed September 20, 2019).

Wald, G. 1945. The spectral sensitivity of the human eye. I. A spectral adaptometer. *J. Opt. Soc. Am.* 35(3):187–196.

Wolska, A. 1998. Psychofizjologia widzenia. Technika świetlna'98. Polski Komitet Oświetleniowy, 135. Warszawa: Stowarzyszenie Elektryków Polskich.

Wong, W. L., X. Su, X. Li et al. 2014. Global prevalence of age-related macular degeneration and disease burden projection for 2020 and 2040: A systematic review and meta-analysis. *Lancet Glob. Health* 2:106–116.

ZUS [Zakład Ubezpieczeń Społecznych]. 2017. Analiza przyczyn absencji chorobowej w latach 2012–2016 [Social Insurance Institution statistics in Poland. Analysis of the causes of sickness absence in 2012–2016]. E. Karczewicz, ed. www.zus.pl/documents/10182/39590/Analiza+przyczyn+absencji+chorobowej+w+latach+2012-2016.pdf/c045c950-143c-4b25-98d7-e0bf5d5dae2e. (accessed September 20, 2019).

3 Selection of Head Protection and Eye and Face Protectors

Central Institute for Labour Protection – National Research Institute

CONTENTS

3.1 GENERAL RULE FOR SELECTION OF PERSONAL PROTECTIVE EQUIPMENT

The general rule for the selection of helmets, eye and face protection is the same as for all other types of personal protective equipment, except that eye protectors must take into account possible visual impairment of users. A summary of several methods described in the literature, which constitute the basis for the proper selection of personal protective equipment, may be proposed by the authors' original, simple method of the CRS triangle (C – protective characteristics; R – hazards/risk occurring at the workplace; P – personal protective equipment whose characteristics correspond to the hazards and risks occurring in a given work environment), the diagram of which is presented in Figure 3.1.

In the diagram presented in Figure 3.1, the vertices of the base of the triangle are the characteristics of personal protective equipment and the hazards and risks occurring at the workplace. This means that for the proper selection of personal protective equipment, basic knowledge of the characteristics of the equipment and of the hazards and risks occurring in the work environment and the resulting risks is necessary.

Conducting a hazard analysis and risk assessment is the first step necessary to perform before proceeding with the selection procedure for personal protective equipment. Any further action is inappropriate without this stage. A detailed hazard analysis and a reliable assessment of the level of risk on its basis make it possible to draw up a list of characteristics that personal protective equipment must have in order for the materials used and the full design of the protective equipment to provide real

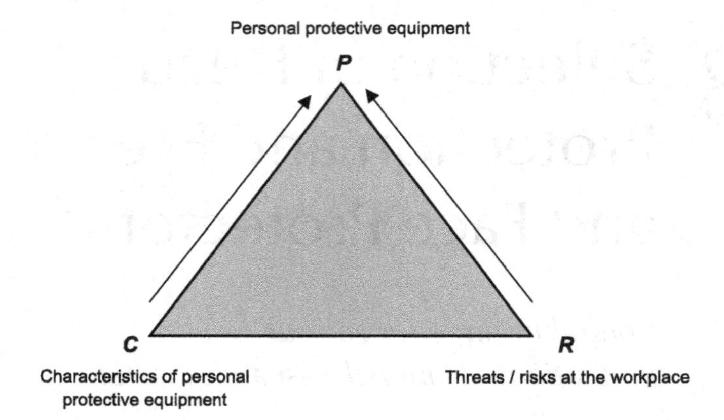

FIGURE 3.1 A diagram illustrating the general principle of selection of personal protective equipment.

protection against the identified hazards, in accordance with the level of risk involved. While drawing up the list of characteristics of personal protective equipment, the individual requirements of the users and the fact that the equipment can be used together with another type of protection (compatibility) should also be considered.

3.2 SELECTION AND USAGE OF SAFETY HELMETS

The first fundamental step in the selection of protective helmets is the assessment of the hazards and environmental conditions occurring at a given workplace. The most important phenomena and factors that need to be identified are:

- The occurrence of falling objects
- The occurrence of stationary objects that pose a risk to hit them with the head
- The action of horizontal forces threatening to squeeze the head transversely
- Atmospheric factors such as negative and positive temperatures, humidity, direct sunlight
- Hot factors such as molten metal splashes, open flame, infrared radiation, etc.
- Aggressive chemicals
- Contact with live metal parts
- Explosive atmosphere
- UV radiation
- Use with other types of personal protective equipment.

The occurrence of mechanical hazards, such as impacts by falling objects and impacts upon dangerous elements of the workplace, determine the type of helmet to be used. Bump caps may be considered if there is no danger from falling objects, only the possibility of superficial injury to the scalp when hitting hard objects. This provides sufficient protection while maintaining comfort. In the event of a risk of impact by falling objects, it is necessary to use at least helmets meeting the requirements of EN 397:2012+A1:2012 [CEN 2012a], ANSI/ISEA Z89.1-2014 [ANSI/ISEA 2014], ISO 3873 1st Edition 1977 [ISO 1977] or AS/NZS 1801:1997 [AS/NZS 1997]. This will also protect the user's head from being hit by stationary objects and structures. If there is a risk of impact on objects with high kinetic energy at the workplace, and in addition the impact can be directed not only to the top, but also to the front, back and sides of the head, it is necessary to use high performance industrial helmets, e.g. meeting the requirements of EN 14052:2012+A1:2012 [CEN 2012b]. The advisability of such a choice is confirmed by the results of tests presented in the articles Baszczyński [2014a] and Korycki [2002], where the influence of the position of the impact point on the helmet shell on its shock-absorbing properties is presented. Typical safety helmets meeting the requirements of EN 397:2012+A1:2012 [CEN 2012a] were tested by measuring the maximum force from the impact of a striker on a headform with the helmet on. It was found that during impacts of a striker with the same kinetic energy, the force applied to the headform with respect to the top point of the helmet shell was about 3 to 5 times smaller than with respect to points located near the lower edge of the shell. This means that helmets of this design are not effective protection against side impacts. The solution to this issue is the use of helmets equipped with protective padding, located between the harness and the shell. In such a helmet design, the shock of the impact energy, as the impact moves to the edge of the shell, is gradually absorbed by the protective padding [Forero Rueda 2009; Gilchrist 1989; Shuaeib 2007].

In industrial practice, there are workstations where there is a risk of lateral deformation, e.g. when carrying loads. In such a situation it is necessary to use a helmet with proven resistance to lateral deformation.

If the work performed at a given workstation poses a risk of the helmet falling off the head due to the employee's position, the helmet should be designed in such a way that the headband fits exactly to the occipital part of the head and is equipped with a chin strap. This problem particularly concerns the simultaneous use of helmets with personal fall protection equipment. In such a situation, during the fall arrest accompanied by swinging motion and collision with the workplace structure, the fact of fastening the helmet under the chin (and thus preventing it from slipping off the head) may determine the health and life of the worker [Baszczyński 2018]. An example of a swinging motion situation when arresting a fall from a height and impacting a flat obstacle with the head is shown in Figure 3.2.

When considering the provision of safety helmets for workers, it is essential to analyse the atmospheric conditions in which they are to be used. This applies primarily to the temperature ranges, i.e. the lowest and highest ambient temperatures. Testing of safety helmets, provided in the report by Baszczyński [2011] and the article by Baszczyński [2014], showed significant sensitivity of their protective properties to variations in temperature at which they are used. Decrease in the shell and harness strength, as well as deterioration of shock absorption properties, manifested by an increased force transmitted to the head of the user during an impact on the helmet of a moving object, constituted the most significant effects observed. The reduction in the strength of the helmet elements in the tests was usually a consequence of initial conditioning at low temperatures (below –20°C). Due to this, during the impact on the helmet mounted on a headform of striker with energy of 49 J (according to the standard EN 397:2012+A1:2012 [CEN 2012a]), there

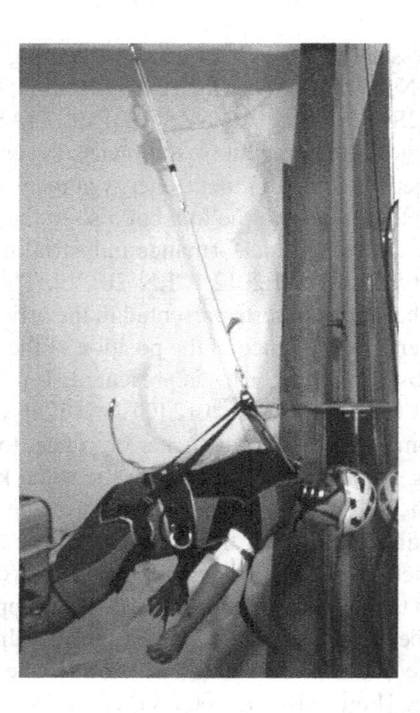

FIGURE 3.2 Head impact on an obstacle in the swinging motion performed while arresting a fall by personal protective equipment.

were cracks in the shells and ruptures of the harness straps and their attachment points joining the harness with seats in the shell. These effects are dangerous from the user's safety point of view, as they constitute a threat of a serious head injury or a rapid increase in the force transmitted to the head and cervical vertebrae. Another dangerous phenomenon, associated with the ambient temperature changes during the use of helmets (mainly its increase), is softening of the materials they are made of. This effect concerns the helmet shell and harness, and is sometimes manifested already in temperatures of 50°C [Baszczyński 2014]. As the material softens, with the same impact energy, the deflection of the shell and the elongation of the harness straps increase. During the testing simulating such a situation, the shell was driven in to a headform, and in real conditions this would be the head of the user. It was associated with an increase in the force transmitted to the headform, which is shown in Figure 3.3.

The tests results presented in the report by Baszczyński, Jachowicz and Jabłońska [2011] and in the article by Baszczyński [2014] showed that with regard to most helmet structures, the helmets' pre-conditioning temperature increases before impact, reducing the shock absorption capacity. The shock absorption capacity is understood here as the value of kinetic energy of the striker impact, in relation to which the value of the force transmitted to the headform reaches 5 kN, according to EN 397:2012+A1:2012 [CEN 2012a]. The aforementioned test results clearly indicate that an accurate examination of thermal conditions concerning the helmet use has a very significant impact on the safety of their users.

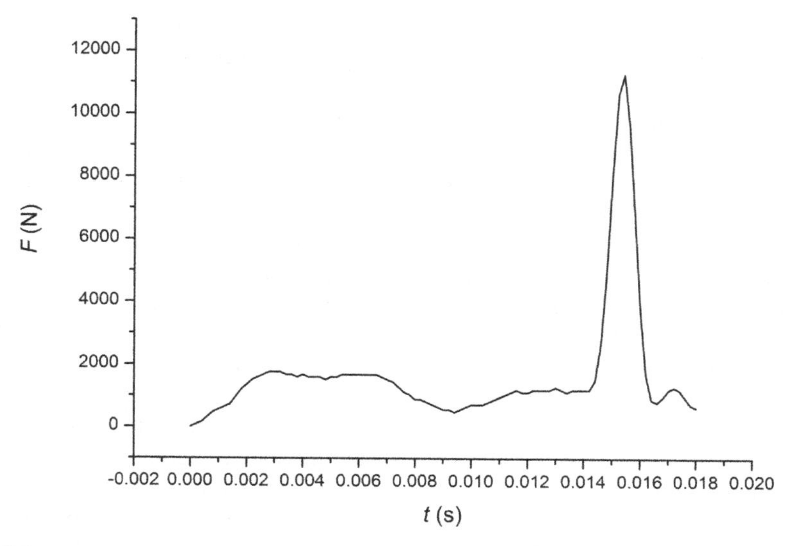

FIGURE 3.3 The course of the force transmitted by the helmet to a headform during an impact of a striker with the energy of 49 J with a characteristic peak, which accompanies the contact of the helmet shell with the headform.

Hot factors in the working environment may also take other forms, such as molten metal splashes, high intensity infrared radiation, open flames, etc. As these may adversely affect the protective properties of the helmets, it is necessary to use designs that guarantee the appropriate properties in this situation. With regard to helmets resistant to molten metal splashes, the shells thereof must be made of material that is not subject to melting, deformation and ignition under the influence of liquid metal particles. In the case of a helmet resistant to intense infrared radiation, the shell thereof must be covered with material reflecting that radiation, e.g. a layer of sprayed metal, which prevents it from overheating. Some structures of infrared resistant helmets are also equipped with a lining that additionally insulates the head of the user.

In industrial conditions, there are workstations where helmets can be exposed to aggressive chemicals. Repeated exposure of the helmet to such substances may result in degradation of the materials they are made of and, consequently, loss of protective properties. Therefore, the decision to use the helmet in such a workstation must be consulted with its manufacturer unless there is adequate information in the instructions for use. Helmets intended for firefighters, meeting the requirements of standard EN 443:2008 [CEN 2008], constitute examples of helmets whose resistance to aggressive chemicals is verified in laboratory tests.

When working in the vicinity of the elements under electric voltage, it is necessary to choose a helmet that has appropriate electrical insulating properties. Such a helmet must protect the user against electric shock, as well as prevent a short circuit, e.g. by touching the helmet shell to two electric wires.

The shells of certain types of protective helmets, due to the material from which they are manufactured, are capable of collecting electrical charges. This property may lead to spark discharges which are dangerous when helmets are used in an explosive atmosphere, e.g. coal mines endangered by methane. These conditions require helmets with antistatic properties, i.e. helmets resistant to electrification and with shells of a low surface resistance, neutralizing the accumulated charge.

The selection of helmets for a specific workplace should also factor in the exposure to UV radiation, which is one of the main factors causing degradation of plastics and, consequently, their ageing. UV radiation sources may be both, natural, i.e. the Sun, and artificial, such as an electric arc during welding. When selecting helmets for such workplaces, attention should be paid to the manufacturers' recommendations in this respect and the helmets with shells sensitive to UV radiation should be avoided, e.g. polythene with no stabilizing additives.

An important aspect of helmet selection is the intention to use them simultaneously with other personal protective equipment, e.g. eye and face protectors, respiratory protective equipment, hearing protectors and protective clothing. In such a situation, helmets must be fitted in such a way that they are compatible with other equipment, i.e. no new hazards are created concerning their user and no protective function of the individual components is impaired. The materials provided by manufacturers of protective equipment, should be the basic source of information in this respect.

The efficiency of the head protection with a helmet is determined not only by the selection, but also by its proper use. Therefore, the following basic principles of using protective helmets were formulated:

- Before use, the helmet should be fitted to the head dimensions by a proper adjustment of headband circumference, wearing height and chin strap length.
- The helmet must be removed of service if it was subject to a severe impact or shows signs of damage.
- The helmet must be removed of service if its life-span has expired, as provided by the manufacturer in the instructions for use.
- The helmet should be stored in the conditions provided by the manufacturer, that do not result in loosing protective parameters.
- Modification of the helmet construction by the user is prohibited.
- Helmet maintenance should be carried out using methods recommended by the manufacturer.

3.3 SELECTION OF EYE AND FACE PROTECTORS

Selection of eye and face protectors is a very broad issue, especially taking into account different risks caused by harmful optical radiation. Depending on the source type, its intensity and radiance spectral distribution, optical protective filters are used for eye protection, the properties of which are described by a number of parameters which determine the characteristics of the filters in detail. Methods of matching optical protective filters to specific applications can be found in the relevant standards. The parameters defining the field of optical protective filters application are defined and determined in accordance with the methodology which is presented in Section 2.2.

When selecting eye and face protectors, individual requirements of users and the compatibility of use with other types of personal protective equipment are very important. Individual requirements of the users of eye and face protectors refer mainly to sight defects (most often refractive defects corrected with spectacles or contact lenses), as well as the use of implants in the form of intraocular lenses (IOLs). These requirements include the ability to fit the face protection in the right position thus ensuring comfortable use and – what becomes more a more important for users – an appropriate presentation.

Compatibility concerns the simultaneous use of eye and face protectors together with other protective measures such as industrial safety helmets, respiratory protective devices (equipment not factory-integrated into eye protection), as well as hearing protectors (in the form of earpieces).

The following study aims to indicate the appropriate approach to the procedure for selecting eye and face protectors, taking into account aspects arising from practice. The aforementioned approach is also intended to help in the interpretation of standards which provide guidelines for the selection of particular types of eye and face protectors. It applies to all standards, regardless of the geographical area of origin. After all, the procedure for the proper selection of eye and face protectors must not depend on whether the requirements of European, Australian or Canadian standards are taken into account. Therefore, the focus given below is the general selection principle, which is illustrated by two examples of eye and face protectors selection.

The first example concerns the selection of eye protectors for welding operations. The second relates to the use of appropriate eye protection while driving vehicles (protection of drivers' eyes). The selection of examples is not accidental. The first example (the welder) relates to the industrial environment and the workplace, which can be taken up exclusively by individuals who have appropriate qualifications, knowledge of hazards involved in the work, and are mandatorily trained in safety (including personal protective equipment). The second example concerns the pursuit of a profession that also requires high qualifications, but in most cases does not expect the worker to have detailed knowledge of hazards and eye protections.

3.3.1 Selection of Eye Protectors for Welding Operations

Despite the automation of welding processes (robot welding), manual welding processes using a wide range of equipment and welding technologies are still common. Steel construction constituting road, marine or industrial infrastructure requires manual welding processes.

The first step in the analysis of hazards concerning welding position is to determine what kind of welding technology is used and the place where the work is carried out. Regardless of the welding technology used, and including welding allied processes, optical radiation constitutes one of the most important harmful factors. Visible (VIS), ultraviolet (UV) and infrared (IR) radiation is emitted during welding and allied processes. Therefore, a welding filter is the basic element for eye protection during welding. The basic division of welding technologies are gas welding processes, arc welding and allied processes. The amount of optical radiation emitted during welding processes can vary greatly. This variation concerns both the proportion of the amount of radiation in the VIS, UV and IR bands as well as power of radiation in the individual spectral bands. It is well known that during arc welding, a very large amount of light is emitted. If the eyes are not adequately protected against the radiation emitted during arc welding, an immediate glare effect will occur, which can lead to eye damage. Gas welding also implies visible radiation emissions, but its amount is much smaller.

The amount of radiation emitted during arc welding depends on the following factors: intensity of the welding current, type of welding technique (e.g. MIG, MAG, TIG) and electrode used, as well as the sort of material of which the components are made. With regard to gas welding, the basic factor influencing the amount of radiation emitted is the gas flow intensity.

A welding filter that would be suitable for eye protection during certain welding processes must therefore suppress the optical radiation in such a way that, depending on the amount of optical radiation emitted, the amount of radiation passing through the filter (on the eye side) is safe. The more intense the welding radiation is, the darker the filter should be. The optical radiation transmission parameters for welding filters are set at a level that factors in the amount of optical radiation that is emitted during certain welding processes. These values are specified in the European standards containing requirements for welding filters EN 169:2002 [CEN 2002], EN 379:2003+A1:2009, [CEN 2003], as well as the new international standards ISO 16321-1 [pr. ISO 16321-1], ISO 16321-2 [pr. ISO 16321-2]. The European standard EN 169:2002 [CEN 2002] also provides guidelines for selecting the filter marking for a particular welding technique, as well as parameters used therein. Simultaneously with writing this monograph, the works aimed at development of ISO 19734 standard (Eye and face protection – Guidance on selection, use and maintenance) are carried out. It will encompass guidelines for the selection of all types of eye and face protectors, including welding filters. Table 3.1 presents a list of markings of welding filters that are used during arc welding and allied processes appropriate to the welding technique used [CEN 2002].

A significant amount of welding dust, gases and fumes may also be emitted during welding. The type of harmful substance mainly depends on the type of material being welded. If the concentrations of welding dust, gases or fumes exceed permissible values, the eye and face protectors should be used together with a respiratory protective device, which is often integrated into welding shields. Regardless of the type of welding, there are mechanical hazards in the form of chips of solid substances (metal, slag). The aforementioned chips are dangerous not only for the eyes but also for the face. For this reason, welding protectors are in the form of welding hand shields, safety helmets or goggles. If there are mechanical hazards in the working environment which force the use of industrial safety helmets, the welding protectors used must be suitable for integration or trouble-free application with the helmet.

During the welding filter selection procedure, factors such as the lighting at the workplace and the state of an employee's sight should be examined. These factors are assessed subjectively. It is very difficult to correlate the dark state of the welding filter with intensity of lighting at the workplace. During welding, the amount of light emitted may be incomparably higher than the amount of light emitted from the lighting at the workstation itself. A similar difficulty is to correlate the state of an employee's sight with the comfort of observing the welded elements through the filter. The use of welding safety helmets, equipped with automatic welding filters

TABLE 3.1
Code Number Corresponding to the Welding Technique Used (Filter Selection Table)

Welding Method	Current Intensity (A)	Code Number (according to EN 169:2002/ISO 16321-2)
Shielded metal arc welding (**E**)	$20 < I \leq 40$	9/W9
	$40 < I \leq 80$	10/W10
	$80 < I \leq 175$	11/W12
	$175 < I \leq 300$	12/W12
	$300 < I \leq 500$	13/W13
	$500 < I$	15/W15
Metal inert gas (**MIG**) welding	$80 < I \leq 100$	10/W10
	$100 < I \leq 175$	11/W11
	$175 < I \leq 250$	12/W12
	$250 < I \leq 350$	13/W13
	$350 < I \leq 500$	14/W14
	$500 < I$	15/W15
Metal inert gas (light) (**MIG-L**) welding	$80 < I \leq 100$	10/W10
	$100 < I \leq 175$	11/W11
	$175 < I \leq 250$	12/W12
	$250 < I \leq 350$	13/W13
	$350 < I \leq 500$	14/W14
	$500 < I$	15/W15
Tungsten inert gas (**TIG**) welding of all metals and alloys	$2 < I \leq 20$	9/W9
	$20 < I \leq 40$	10/W10
	$40 < I \leq 100$	11/W11
	$100 < I \leq 175$	12/W12
	$175 < I \leq 250$	13/W13
	$250 < I \leq 400$	14/W14
Electro-air gouging (**EG**)	$125 < I \leq 175$	10/W10
	$175 < I \leq 225$	11/W11
	$225 < I \leq 275$	12/W12
	$275 < I \leq 350$	13/W13
	$350 < I \leq 450$	14/W14
	$450 < I$	15/W15
Plasma stream cutting (**PSC**)	$60 < I \leq 150$	11/W11
	$150 < I \leq 250$	12/W12
	$250 < I \leq 400$	13/W13

(Continued)

TABLE 3.1 (*Continued*)
Code Number Corresponding to the Welding Technique Used (Filter Selection Table)

Welding Method	Current Intensity (A)	Code Number (according to EN 169:2002/ISO 16321-2)
Microplasma welding (MW)	0,1 < I ≤ 0,4	3/W3
	0,4 < I ≤ 0,6	4/W4
	1 < I ≤ 2,5	5/W5
	2,5 < I ≤ 5	6/W6
	5 < I ≤ 10	7/W7
	10 < I ≤ 15	8/W8
	15 < I ≤ 30	9/W9
	30 < I ≤ 60	10/W10
	20 < I ≤ 40	11/W11
	60 < I ≤ 125	12/W12
	125 < I ≤ 225	13/W13
	225 < I ≤ 450	14/W14
	450 < I	15/W15
MAG	40 < I ≤ 80	10/W10
	80 < I ≤ 125	11/W11
	125 < I ≤ 175	12/W12
	175 < I ≤ 300	13/W13
	300 < I ≤ 450	14/W14
	450 < I	16/W16

with the possibility of adjusting a darkness degree, ensures its individual adaptation to the welding technique applied, external lighting conditions and individual visual impressions of the employee.

With reference to the general rule for the selection of personal protective equipment (see Figure 3.1), Table 3.2 presents a list of eye and face hazards during arc welding, characteristics of protectors and type of eye and face protectors used.

Due to the relatively large space between the face and inner surface of welding hand shields and safety helmets used for arc welding, for most spectacles frames and welding shields, there is no problem concerning the simultaneous use of corrective spectacles. Nevertheless, the aforementioned problem does concern hinged goggles used for gas welding. This type of goggles adheres directly to the face and is held by a strap surrounding the head, making it impossible to use corrective spectacles at the same time. One of the methods of ensuring proper vision correction – if required – is to place special holders inside the goggles to fix the corrective lenses.

Gas welding requires much brighter filters than most arc welding technologies. Nevertheless, there is still a significant risk of harmful optical radiation. Table 3.3 shows a list of the required welding filters code numbers used in gas welding, depending on the gas flow intensity q [CEN 2002].

TABLE 3.2
List of Eye and Face Hazards During Arc Welding, Characteristics of Protectors and Type of Eye and Face Protector Used

Hazards	Protector Characteristics	Type of Protectors
Optical radiation within UV-VIS-IR beams.	Suppressing radiation emission to the level safe for eyes.	Welding filter (passive or automatic). Filters – depending on the technology used, marked[a] from 9/W9 to 16/W16.
Chips of solid substances.	Eye protectors and face shield.	Welding hand shield (with passive welding filter) or welding safety helmet (with passive or automatic welding filter).
Gases, vapours and fumes.	Filtration of gases, vapours or fumes to a level lower than the occupational exposure limit (OEL).	Welding safety helmet integrated into airways protective equipment (with automatic welding filter).
Impacts, falling objects, etc.	Head protection provision.	Welding safety helmet integrated into industrial safety helmet (with passive or automatic welding filter).

[a] According to EN169:2002/ISO 16321-2.

TABLE 3.3
Code Number Corresponding to the Gas Welding Technique Used (Filter Selection Table)

Code Number[a]	Gas Flow Intensity q (litre/hour)
4/W4	$q \leq 70$
5/W5	$70 < q \leq 200$
6/W7	$200 < q \leq 800$
7/W8	$q > 800$

[a] According to EN169:2002/ISO 16321-2.

When preparing a list of hazards to the eyes and face during gas welding, it should be noted that despite the fact that the risks are similar in nature, the results thereof are not as great, as in the case of arc welding. There is no doubt that during gas welding there is a risk of chips of solid substances. The nature of these hazards requires eye protectors, without the absolute necessity of a full face shield. For this reason, simple eye protectors in the form of goggles are enough during gas welding.

TABLE 3.4
List of Hazards to the Eyes and Face During Gas Welding, Characteristics of Eye Protectors and Type of Eye and Face Protector Used

Hazards	Protectors Characteristics	Type of Protectors
Optical radiation within UV-VIS-IR beams.	Suppressing radiation emission to the level safe for eyes.	Welding filter (passive or automatic). Filters – depending on the technology used, marked[a] from 1,2/W1,2 to 8/W8.
Chips of solid substances.	Eyes protectors. In specific cases, all face shield.	Hinged goggles for welder. In specific cases, safety helmet with automatic welding filter.
Gases, vapours and fumes.	Filtration of gases, vapours or fumes to a level lower than the occupational exposure limit (OEL).	If the OEL's value is exceeded, use of respiratory protective device that can be used with protective goggles should be applied. If a welding safety helmet is used, it should be integrated into airway protective equipment (with automatic welding filter).
Impacts, falling objects, etc.	Head protection provision.	The use of hinged goggles permits trouble-free use of any type of industrial safety helmet. If a welding safety helmet is used, it must be integrated into industrial safety helmet (with automatic welding filter).

[a] According to EN169:2002/ISO 16321-2.

In particular cases, where the specificity of the technological process performed results in hazards in the form of chips which are also dangerous for the face, full face protection would be required. In such a case, a safety helmet with automatic welding filter will be an excellent eye and face protector, which may darken to the level required during gas welding. Table 3.4 shows a list of hazards to the eyes and face during gas welding, characteristics of eye protectors and type of eye and face protector used.

Carrying out welding works often requires participation of persons assisting in the welder's tasks (in particular during arc welding). These individuals, called "welder's assistants", usually stay away from the source constituting a danger to the eyes and face. They are not exposed to direct observation of the welding process. Nevertheless their eyes are subject to harmful optical radiation. In order to ensure safety, such persons should wear safety goggles for the welder's assistant. The dark state of welding filters used in such type of spectacles is generally much lower than those used by the welder. It is comprised in the marking range from 1,2/W1,2 to 5/W5 [CEN 2003, pr. ISO 16321-2].

To sum up the aforementioned analysis of the problem concerning selection of eye and face protectors during welding works, it should be stated that:

- During arc welding and in allied processes, the most important factor determining the dark state of protective filters is current intensity. Depending on the technology used, filters marked from 9/W9 to 16/W16 are used.
- The basic type of eye and face protectors in arc welding and allied processes is a welding hand shield or safety helmet. Currently, welding safety helmets with automatic welding filters are commonly used.
- During gas welding and in allied processes, the most important factor determining the dark state of protective filters is the intensity of gas flowing through the burner. Filters marked from 1,2/W1,2 to 8/W8 are used.
- The basic type of eye protectors in gas welding is hinged welding goggles.
- In most cases, the use of welding hand shields and safety helmets does not cause a collision in the use of corrective spectacles.
- When welding goggles are used, it is not possible to use corrective spectacles at the same time. Persons requiring sight correction must use welding goggles designed to incorporate corrective lenses.
- Persons assisting in welding works – mainly in arc welding – should protect the eyes from harmful optical radiation. Welding filters fitted in these spectacles are marked from 1,2/W1,2 to 5/W5.

3.3.2 Selection of Eye Protectors for Drivers

The choice of eye protection for drivers concerns both road and rail infrastructure workers, including car drivers (cars, trucks), motorcyclists, as well as cyclists, and increasingly common urban users of private means of transport, such as electric scooters.

The driver's eyes are exposed to optical radiation, and in particular to intense visible radiation that can cause glare. The glare results in a temporary (up to several seconds) loss of vision, which can lead to an unfortunate accident. All persons operating motor vehicles must be aware of the hazards caused by glare, i.e. usually short-term, excessive solar radiation or radiation from headlights of other vehicles.

The risks to the eyes and face of road users depend to a large extent on whether the vehicles are equipped with a windscreen, which naturally provides protection against external factors such as dust, chips of solid substances, liquid drops (rain), etc. Otherwise, drivers are directly exposed to the above mentioned factors. When driving a cab vehicle, the glare effect can be reduced or eliminated by adjusting the glareshields which are fitted to the vehicle. If the vehicle does not have a cab, the only protection against glare is through the use of filters fitted to the safety spectacles. As these filters mainly protect against solar radiation, they are referred to as sunglare filters, commonly called sunglasses. They protect the eyes from excessive sunlight, reduce eye fatigue and, in special cases, improve visual stimuli. The filter darkening used in sunglasses is selected according to the ambient light conditions and the individual sensitivity of the user's eye to the glare, including long-term use without additional, undesired fatigue effect.

Professional selection of sunglasses, both in everyday and occupational use, should, first of all, include an estimate of how much light must be absorbed by the filters in order to ensure maximum safety and comfort of vision and possible maintenance of vision-enhancing properties. The minimum value of the luminous transmittance factor for sunglasses suitable for driving is 8%, as specified by the requirements of PN-EN ISO 12311:2013 [CEN 2013]. Employers, as well as the users of sunglasses themselves, should also be aware that the car window can also block out some of the optical radiation. Therefore, it should be clearly stated that the maintenance of the adequate level of protection and quality of observation has an additive effect on the individual elements of the system that are in front of the eye. It includes, among others, spectacles with corrective effect and sunscreen in the form of e.g. a shade cap and a car window. All these elements can block a certain amount of light reaching the eye. Under the main assumption that the level of filtration should not exceed 8% and the acceptable level of transmittance of car windows is 75%, it is possible to estimate the safe maximum level of sunglasses filters used by drivers.

In practice, in addition to including the tinting of car windows and the transmittance of spectacles with corrective effect, the dynamically changing lighting conditions and the individual sensitivity of the eye to glare must be taken into account. Therefore, the decision to choose the optimal (compliant with standards) darkening of the sunglasses filters is ultimately made by the users themselves.

Sunlight filters, apart from protection against glare, can improve the observation due to, among others, the phenomenon of polarization. Such a property is obtained by means of a suitably selected filter colour for road conditions or by using special glare shields. As a result of polarization, the effects in the form of so-called mirages, which are formed on the road in conditions of high temperature and extremely hot air, are eliminated. The properly selected colour of the filters increases the visual acuity and contrast of observations in mist or dusk conditions. Yellow filters, so-called *blue blocker*, are commonly used. The adequate selection of the filter colour also helps in the proper recognition of traffic lights. The use of sunglasses that do not obstruct the recognition of traffic lights is particularly important for people operating rail vehicles. In these conditions, it is necessary to recognize lights in a complex railway signalling system that uses a variety of colours in different graphic configurations. Anti-reflection coating – also made with interference technology – reduces the risk of glare from vehicle headlights.

Because of the risk of optical radiation (glare, as well as lighting conditions), the characteristics of filters used in driver's eye protection are as follows: darkening (determined by the value of luminous transmittance), colour, polarization and properties related to the elimination of unwanted reflections (anti-reflection coating).

The characteristic features of the construction of spectacles used in vehicles without a cab are the following: provision of an adequate field of vision and lateral protection, stability on the head of the user (protection against falling off the head) and adequate mechanical strength. The latter feature is particularly important for visors used in motorcycle helmets, etc. The constructions of this type of helmet, including visor constructions, are of very high mechanical strength due to extreme head hazards.

Selection of eye protection for professional drivers, regardless of the use conditions, requires special attention to the occurrence of visual impairment (to the extent permitted). This mainly concerns refractive errors, colour recognition and cataracts, and a dark adaptation defect. These defects often occur simultaneously.

In the case of refractive errors, it is necessary to use spectacles and observe the mandatory periodic vision checks. The lenses used in spectacles can have different colours or change the luminous transmittance automatically (photochromic effect). Drivers wearing spectacles with corrective effect to protect themselves from glare use clip-on sunglasses or sunglasses with individually selected refractive parameters.

When using clip-on sunglasses, the darkening and colour of the filter should be chosen in such a way that the luminous transmittance in the prescription spectacles – clip-on – windscreen system is not lower than 8%. The other optical parameters, which are decisive for undisturbed colour recognition, should also be at an appropriate level in this configuration. There are no laboratory tests on luminous transmittance of optical systems consisting of prescription spectacles, sunglasses, clip-on sunglasses and car windscreens. The characteristics determining the optical radiation transmission properties are determined only with respect to clip-on sunglasses. Drivers using prescription spectacles with clip-on sunglasses should be aware of the total transmittance as defined in the standard. The darkening of the already relatively dark spectacle lenses may cause the transmittance factor of the entire prescription spectacle system – corrective – clip-on system to be too low. However, most drivers handle the above described problem very well. The drivers intuitively adjust the darkening of their spectacles to the conditions on the road and their vision. This does not mean, however, that the issue of the dark state of filter selection used by drivers should be reduced solely to the subjective assessment of the driver himself. This assessment is extremely important. The value of luminous transmittance of clip-on sunglasses used by drivers as glare protection can be estimated with the knowledge of the luminous transmittance of prescription spectacles and car windscreen. The luminous transmittance of the system: prescription spectacles – clip-on – windscreen is equal to the product of the luminous transmittance factors, as expressed in formula 3.1:

$$\tau_{F\_O+N+S} = \tau_{F\_O} \cdot \tau_{F\_N} \cdot \tau_{F\_S}, \tag{3.1}$$

where:

τ_{F_O+N+S} – luminous transmittance factor of the system: prescription spectacles – clip-on – windscreen

τ_{F_O} – luminous transmittance factor of the prescription spectacles

τ_{F_N} – luminous transmittance factor of the clip-on sunglasses

τ_{F_S} – luminous transmittance factor of the windscreen.

The formula (3.1) allows easy calculation of the luminous transmittance of the clip-on sunglasses knowing the values of the other luminous transmittance factors. If the transmittance factor of the whole system must not exceed 8% (0.08), and the estimated value of transmittance factors of prescription spectacles itself and windscreen are at the level of 75% (0.75), the value of luminous transmittance of the cover plate, determined by the formula 3.1, amounts to approximately 14% (0.14). The accepted values of luminous transmittance with regard to the prescription spectacles and windscreen are at minimal transmittance levels that are characteristic of spectacle lenses without filters and windscreens. The value of transmittance factor of sun-protective cover plate, determined by the formula 3.1, indicates the third category of the filter protecting against glare. For this category, the minimal value of luminous transmittance is set at 8%, with the maximum value at 18%.

The above calculations may be used as a basis for determining the minimum value of the clip-on sunglasses luminous transmittance factor, so that, taking into account the light attenuation of the spectacles and the lens itself, the darkening of the clip-on is not too high. Much more complicated calculations, requiring detailed data on the spectral transmittance characteristics of all components of the prescription spectacle – cap-on sunglasses – lens system, are necessary to determine the factors characterizing potential colour recognition interference.

In addition to the spectacles with the corrective effect, drivers use progressive and polarized spectacles. Progressive spectacles users must take into account the specificity of vision with this type of spectacles. If the user is not adapted to use progressive spectacles, they may, for example, lose the sharpness of the images they perceive in mirrors or on vehicle dashboard displays. The polarization effect, which is extremely beneficial due to the elimination of artefacts appearing on the road in certain weather conditions, may in turn prevent proper observation of some liquid crystal displays. The use of progressive spectacles for driving requires an appropriate adaptation period. When using polarizing spectacles, a preliminary check should be made to ensure that the polarization effect does not interfere with the observations necessary to operate the vehicle (e.g. liquid crystal displays of navigation systems, etc.).

REFERENCES

ANSI/ISEA [American National Standards Institute (ANSI)/International Safety Equipment Association (ISEA)]. 2014. Industrial head protection ANSI/ISEA Z89.1-2014. Washington, DC.

AS/NZS [Standards Australia, and Standards New Zealand]. 1997. Occupational protective helmets AS/NZS 1801:1997. Homebush, Australia and Wellington, New Zealand: AS/NZS.

Baszczyński, K. 2014a. Przemysłowe hełmy ochronne a zabezpieczenie głowy przed uderzeniami bocznymi [Industrial safety helmets and protection of the head against side impacts]. *Bezpieczeństwo Pracy – Nauka i Praktyka* 5(370):10–13.

Baszczyński, K. 2014b. The effect of temperature on the capability of industrial protective helmets to absorb impact energy. *Eng. Fail. Anal.* 46:1–8.

Baszczyński, K. 2018. Analiza zagrożeń uszkodzeń głowy pracownika podczas powstrzymywania spadania z wysokości [Risk analysis of employee head injury during a fall arrest]. *Bezpieczeństwo Pracy – Nauka i Praktyka* 2(557):20–23.

Baszczyński, K., M. Jachowicz, and A. Jabłońska. 2011. Sprawozdanie z projektu 03.A.09. Opracowanie metody badania hełmów ochronnych w zakresie skuteczności ochrony przed uderzeniami poruszających się obiektów. [Project report 03.A.09. Development of the test method, for helmets, of effectiveness of protection against impacts of moving objects]. Warszawa: Centralny Instytut Ochrony Pracy – Państwowy Instytut Badawczy.

CEN [European Committee for Standardization]. 2002. Personal eye-protection – Filters for welding and related techniques – Transmittance requirements and recommended use EN 169:2002. Brussels, Belgium: CEN.

CEN [European Committee for Standardization]. 2003. Personal eye-protection – Automatic welding filters EN 379: 2003+A1:2009:2001. Brussels, Belgium: CEN.

CEN [European Committee for Standardization]. 2008. Helmets for fire fighting in buildings and other structures EN 443:2008. Brussels, Belgium: CEN.

CEN [European Committee for Standardization]. 2012a. Industrial safety helmets. EN 397:2012+A1:2012. Brussels, Belgium: CEN.

CEN [European Committee for Standardization]. 2012b. High performance industrial helmets EN 14052:2012+A1:2012. Brussels, Belgium: CEN.

CEN [European Committee for Standardization]. 2013. Personal protective equipment – Test methods for sunglasses and related eyewear EN ISO 12311:2013. Brussels, Belgium: CEN.

Forero Rueda, M. A., L. Cui, and M. D. Gilchrist. 2009. Optimisation of energy absorbing liner for equestrian helmets. Part I: Layered foam liner. *Mater. Des.* 30:3405–3413.

Gilchrist, A., and N. J. Mills. 1989. Improved side, front and back impact protection for industrial helmets. School of Metallurgy and Materials. University of Birmingham. Health & Safety Executive contract research report no. 13/1989.

ISO [International Organization for Standardization]. 1977. Industrial safety helmets. ISO 3873 1st Edition 1977. Geneva, Switzerland: ISO.

Korycki, R. 2002. The damping of off-central impact for selected industrial safety helmets used in Poland. *Int. J. Occup. Saf. Ergo.* 1(8):51–70.

Project ISO 16321-1. Eye and face protection for occupational use – Part 1: General requirements.

Project ISO 16321-2. Eye and face protection for occupational use – Part 2: Additional requirements for protectors used during welding and related techniques.

Shuaeib, F. M., A. M. S. Hamouda, S. V. Wong, R. S. Radin Umar, and M. M. H. Megat Ahmed, 2007. A new motorcycle helmet liner material: The finite element simulation and design of experiment optimization. *Mater. Des.* 28:182–195.

4 Compatibility of Safety Helmets with Eye and Face Protectors

Central Institute for Labour Protection – National Research Institute

Serious hazards from mechanical factors arise in many workplaces in industries such as construction, mining, energy, metallurgy, forestry, etc. Mechanical factors are understood here as the contact of the human body with material objects of different masses, shapes and speeds. Such hazards are particularly important if they affect the upper part of the head and face, especially the eyes. Taking into account the variety of activities performed at worksites in the previously mentioned industries, it is not always possible to eliminate mechanical risks by taking appropriate organizational actions or using collective protective equipment. In many cases, the only possibility to protect workers is the simultaneous use of safety helmets with eye and face protection, such as visors, spectacles or safety goggles [Baszczyński 2014]. So far, this issue has been neglected when placing new types of personal protective equipment on the market. Their instructions for use do not indicate the possibility of simultaneous use with other protective equipment. As a result, in practice, situations of using mismatched equipment can be observed, which impairs their protective properties. One of the few exceptions in this respect is the information contained on the website of the Institut fuer Arbeitsschutz der Deutschen Gesetzlichen Unfallversicherung (IFA). Normative documents related to the equipment under consideration often contain requirements relating to the need to ensure compatibility, without specifying how it should be verified.

The problem of compatibility of helmets with eye and face protective equipment concerns two basic aspects:

- Maintaining the protective parameters of the set of personal protective equipment, including not posing any additional risk, and
- Comfort of use.

The following types of equipment should be considered when examining the possibility of simultaneous use of helmets with eye and face protection:

- Safety spectacles,
- Safety goggles,
- Visors with their own headgear or designed to be attached to a helmet.

Visors fixed to helmets are designed to be used together with this equipment. They can be mounted with the use of special attachments placed in the holes of the shell or with a clamping ring surrounding the shell of the helmet above its lower edge. The correctness of the visor's design and its connection to the helmet, which are prerequisites for providing the user with an adequate level of protection, is tested in accordance with specific methods, such as those contained in EN 168:2001 [CEN 2001a]. The assessment criteria in this respect are contained in the standard EN 166:2001 [CEN 2001b]. Most visors with their own headgear are incompatible with protective helmets in practice. This is due to the design of a typical headgear and its adjustment mechanisms, which make it impossible to fit under the headband, harness and shell of the helmet properly.

Helmets can also be used with safety spectacles and goggles. Examples of such combination are shown in Figure 4.1.

The first basic question that arises in the context of wearing a helmet and safety goggles/spectacles at the same time is whether they do not interfere with each other, making it impossible to wear properly. The best method of assessment in this respect, presented in the report by Baszczyński [2011], is to use standard headforms (meeting the requirements of EN 168:2001 [CEN 2001a] or EN 960:2006 [CEN 2006]), to which protective helmets and goggles/spectacles are applied separately. This equipment shall be photographed separately in at least two planes or scanned with a 3D scanner. The images or scans are then superimposed, resulting in an image of the mutual position and potential spatial conflict of the equipment. An example of the result of such tests is shown in Figure 4.2.

The conducted tests by Baszczyński [2011] have shown that in the vast majority of cases the safety spectacles do not come into contact with the shell and the headband of the helmet, which means that they can be used simultaneously when worn under regular conditions. In the case of the various types of safety goggles tested, the area occupied by them on the headform partly coincided with the area occupied by

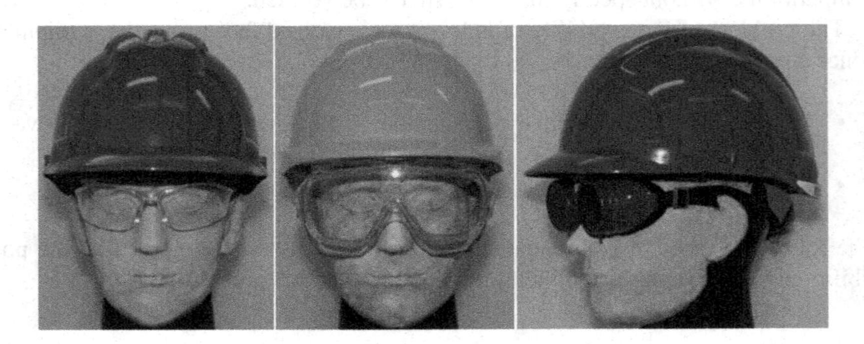

FIGURE 4.1 Example of simultaneous use of helmets and safety spectacles and goggles. (From Baszczyński, K., *Saf. Work Sci. Prac.*, 11, 18–21, 2014.)

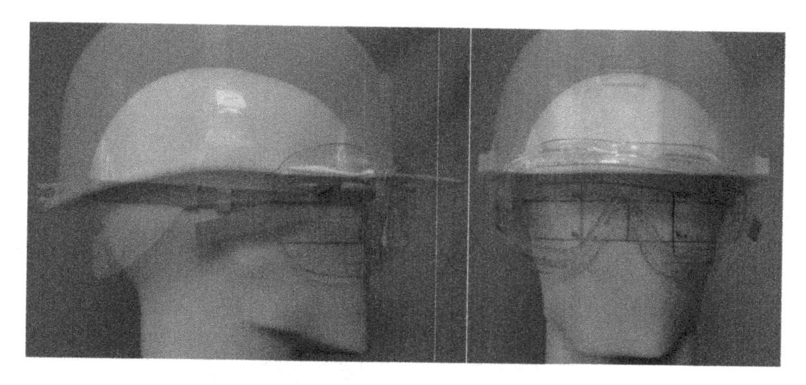

FIGURE 4.2 An example of superimposing images of a helmet and protective goggles worn on the same headform.

the headband of the helmet. Protective goggles also came into contact with helmet peaks. This means that in such cases they cannot be used simultaneously with helmets. In conclusion, it can be said that in most cases protective goggles are not compatible with helmets.

In the case of safety spectacles and goggles, which do not come into contact with the helmet during regular wear, the question arises as to what happens when a strong vertical impact on its shell occurs. The test results presented in the report by Baszczyński [2011] showed that the falling weight impact on the helmet causes deformation of the shell and vertical displacement of the helmet. If eye and face protection is used at the same time, the impact on the shell may be transmitted to the eyes and face, posing a danger to the user's face, e.g. to the nose. In order to explain these phenomena, the tests presented in the article by Baszczyński [2018] were conducted. They were performed on sets of helmets and protective spectacles and goggles as shown in Figures 4.3 and 4.4.

The essence of the conducted tests was to determine the phenomena accompanying falling weight impact on safety helmets mounted on a headform in conjunction with protective spectacles/goggles. For this purpose, the following mechanical parameters were selected:

- Maximum force acting on the helmet upon falling weight impact,
- The maximum value of the helmet edge displacement and its deformation during impact,
- Maximum force exerted by the spectacles/goggles on the headform upon falling weight impact on the helmet,
- Displacement and damage to spectacles/goggles.

In the presented method the helmet is hit by a vertically falling striker with a mass of m = 5 kg and kinetic energy E_k = 49 J, in accordance with the requirements of EN 397:2012 + A1:2012 [CEN 2012]. During the tests, the acceleration of the striker and of the protective spectacles/goggles was measured using an electronic measuring system. Based on the known accelerations, the values of forces acting between the helmet shell

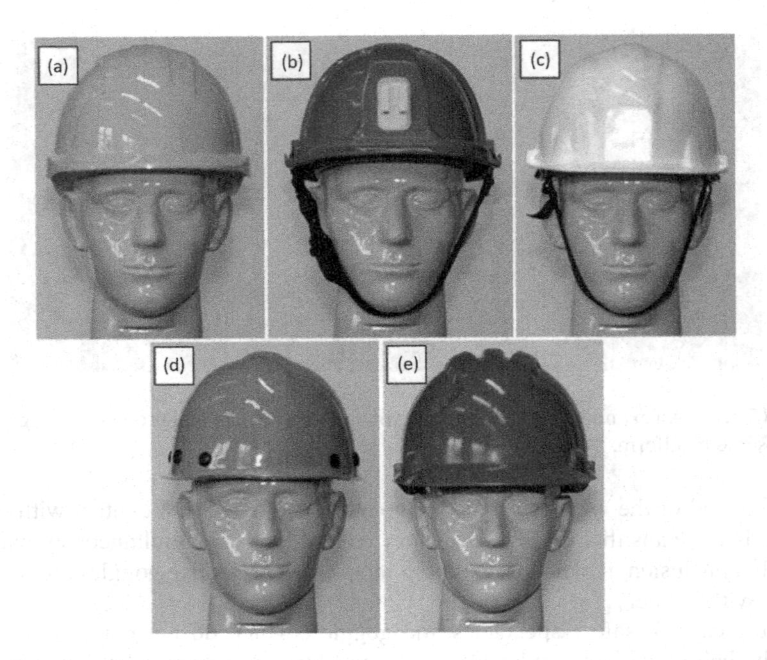

FIGURE 4.3 Safety helmets used in the tests. Note: A–E symbols used in Figures 4.6 through 4.9. (From Baszczyński, K., *Int. J. Occup. Saf. Ergon.*, 24, 171–180, 2018.)

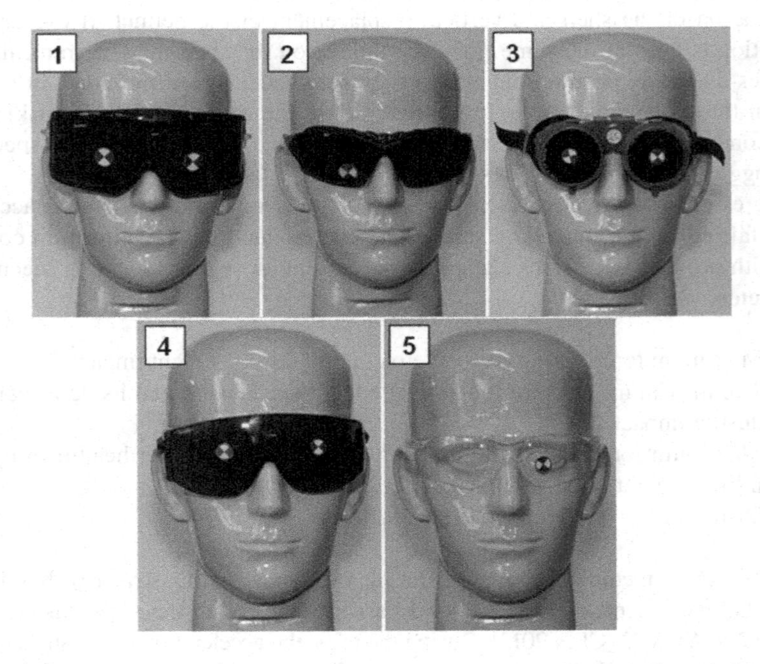

FIGURE 4.4 Safety spectacles and goggles used in the tests. Note: 1–5 symbols used in Figures 4.6, 4.8 and 4.9. (From Baszczyński, K., *Int. J. Occup. Saf. Ergon.*, 24, 171–180, 2018.)

and the striker, as well as between spectacles/goggles and the headform were calculated. The displacement of the helmet shell rim and spectacles/goggles was recorded using a digital high-speed camera set to 2000 frames/second. Detailed description of the measuring apparatus and test methods is presented in the article by Baszczyński [2018]. Sample images from the digital high-speed camera showing the displacement of the striker, helmet rim, spectacles and helmet deformation during impact are shown in Figure 4.5.

According to the performed tests, the helmet shell was subjected to forces with maximum values, shown in the diagram in Figure 4.6.

These results indicate that all helmets complied with the shock absorption requirement of EN397:2012 + A1:2012 [CEN 2012], which means that the maximum force transmitted during the impact on the headform did not exceed 5 kN. The values of recorded maximum displacements and deformations of helmets are shown in Figure 4.7.

This figure shows that the downward movement of the rim, depending on the type of helmet, ranged from 11 to 23 mm. This means that for spectacles/goggles that

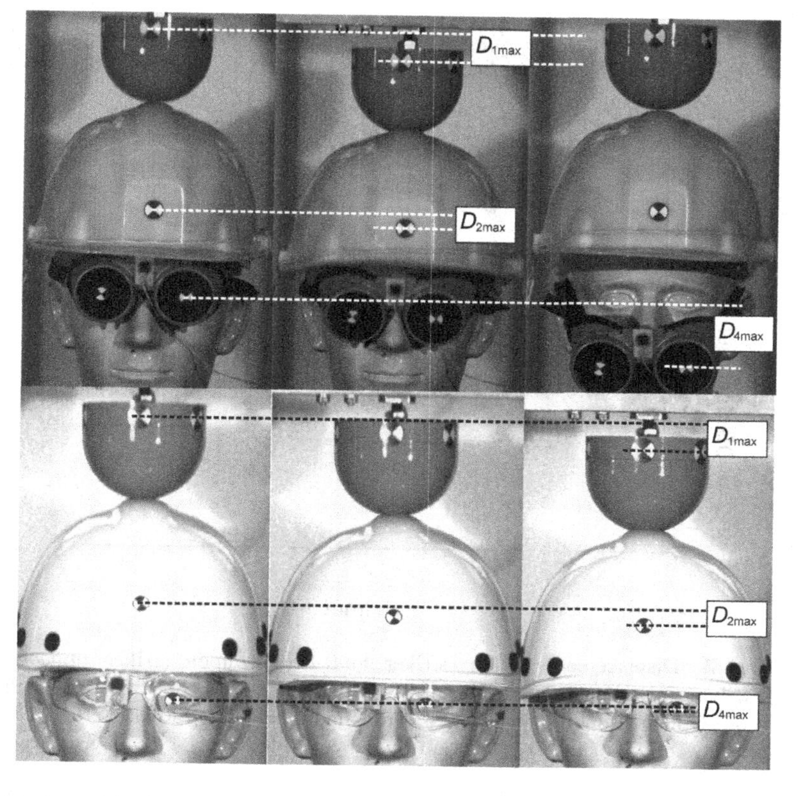

FIGURE 4.5 Displacement of the striker, helmet and spectacles/goggles upon impact on the top of the helmet. Note: D_{1max} = striker displacement upon impact on the helmet; D_{2max} = displacement of the bottom part of the helmet upon impact; D_{4max} = displacement of protective spectacles/goggles resulting from helmet impact. (From Baszczyński, K., *Int. J. Occup. Saf. Ergon.*, 24, 171–180, 2018.)

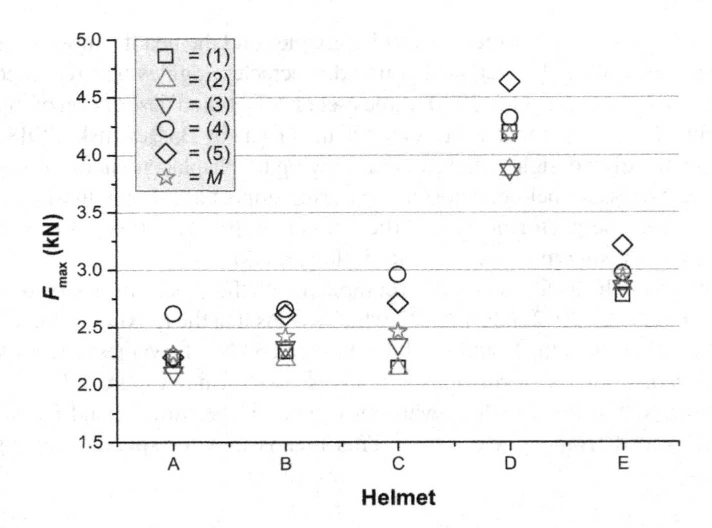

FIGURE 4.6 Maximum forces, F_{max}, acting on the helmet upon striker impact. (From Baszczyński, K., *Int. J. Occup. Saf. Ergon.*, 24, 171–180, 2018.)

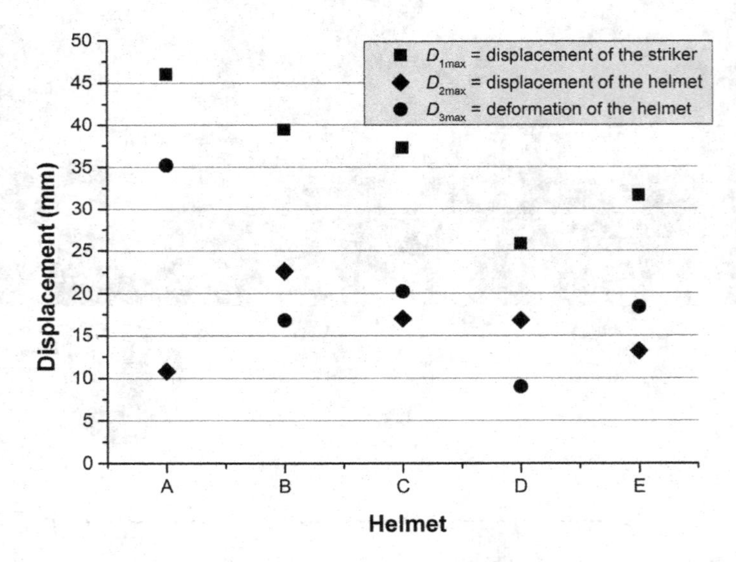

FIGURE 4.7 Displacement of the striker and helmet upon striker impact. Note: D_{1max} = striker displacement upon impact on the helmet; D_{2max} = displacement of the bottom part of the helmet upon impact; D_{3max} = maximum helmet deformation, defined as the difference between D_{1max} and D_{2max}. (From Baszczyński, K., *Int. J. Occup. Saf. Ergon.*, 24, 171–180, 2018.)

come into contact with the helmet shell during regular wear, a vertical force will act on the spectacles/goggles during impacts.

This is confirmed by the results shown in Figures 4.8 and 4.9. The maximum displacement values of the helmets with spectacles/goggles reach approximately 22 mm

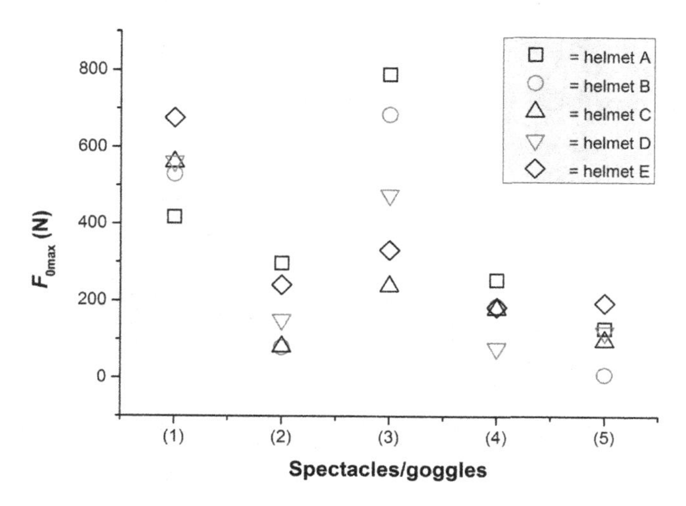

FIGURE 4.8 Maximum forces, F_{0max}, exerted by spectacles/goggles on the headform upon striker impact. (From Baszczyński, K., *Int. J. Occup. Saf. Ergon.*, 24, 171–180, 2018.)

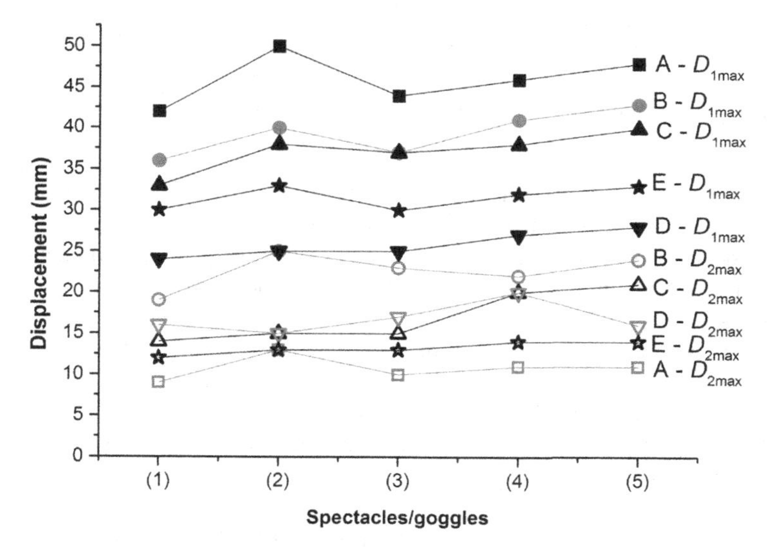

FIGURE 4.9 Displacement of the striker and helmet depending on the type of safety helmet and spectacles/goggles. Note: D_{1max} = striker displacement upon impact on the helmet; D_{2max} = displacement of the bottom part of the helmet upon impact. (From Baszczyński, K., *Int. J. Occup. Saf. Ergon.*, 24, 171–180, 2018.)

and the forces between the spectacles/goggles and the headform, mainly in the nose area, reach approximately 800 N. Given that the identified forces exert on the user's face, this can be considered a serious risk. The results presented indicate that the combined use of eye and face protection products with safety helmets must be preceded by laboratory tests to verify their compatibility under conditions of falling weight impact.

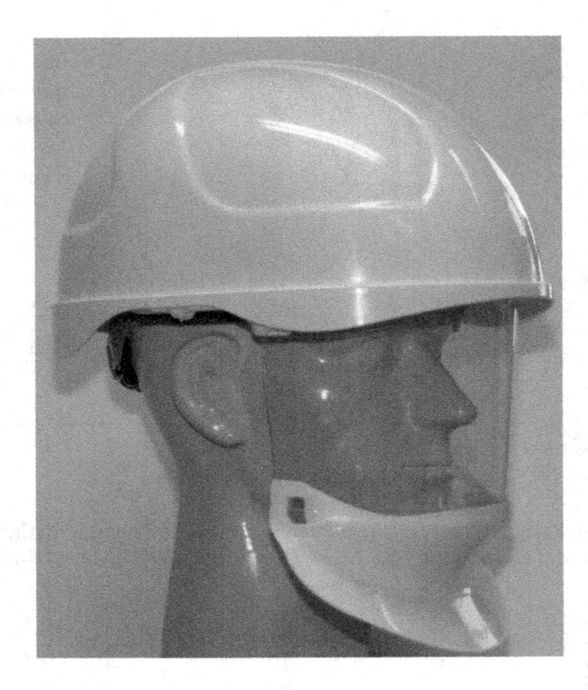

FIGURE 4.10 Example of a helmet with an integrated visor.

Therefore, it can be argued that the best solution to this problem is to provide the user with integrated structures. Examples of such equipment are helmets with integrated visors that can be lowered and lifted. When lifted, these shields typically extend above the helmet shell or hide between the shell and the harness, as shown in Figure 4.10.

Irrespective of the information provided by the manufacturers of protective equipment, the users should check the compatibility of helmets and safety goggles/spectacles by themselves. The essence of such a check is to take into account the individual characteristics of the user, e.g. head dimensions. Self-assessment should include the following steps:

- According to the instructions for use provided by the manufacturers of the equipment, the helmet and safety spectacles or goggles must be adjusted and put on.
- If the helmet is equipped with a chin strap, it should be placed underneath the chin and its length should be adjusted so that it performs its function properly.
- Then make sure that the chin strap does not interfere with the use of spectacles/goggles. To do this, move your head vigorously, observing that the chin strap does not press on the spectacle/goggle components (at the points shown in Figure 4.11) causing them to move, fall or put unacceptable pressure.

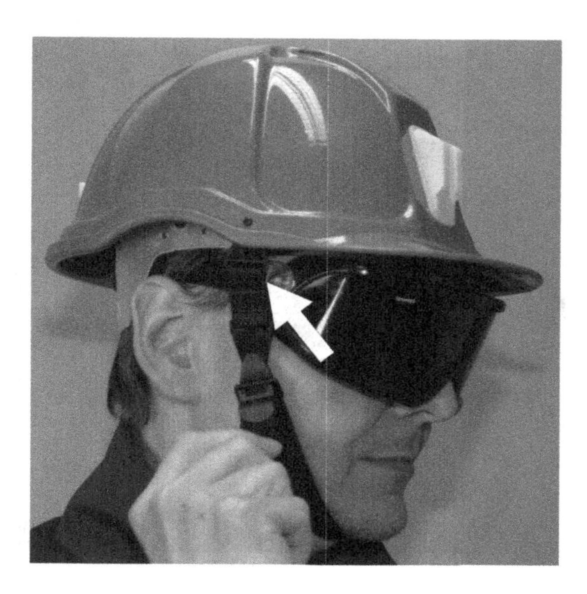

FIGURE 4.11 Control point for chin strap contact with safety spectacles/goggles.

- Check that the edge of the helmet, e.g., the bottom of the peak, does not press on the safety spectacles/goggles. For this, use your fingertips to check if there is a gap between the helmet and the spectacles/goggles to allow you to put on and fit your protective equipment freely.
- Check the behaviour of the helmet – safety spectacles/goggles set by simulating a light falling weight impact. For this purpose, following Figure 4.12, hit the top of the helmet with a slightly open hand and observe whether this will transfer to the spectacles/goggles and cause unacceptable pressure on the nose or other parts of the face. Distinct pressure is a disqualifying outcome.
- When the protective set is fitted, check that the helmet harness (headband and straps) does not come into contact with the spectacle/goggle elements and cause unacceptable pressure or other discomfort.
- In the case of visors mounted on helmet shells, it is necessary to check that the visor does not detach from the helmet during vigorous head movements and bows, and that the helmet does not tend to move or fall. In case of a negative result, the helmet should be considered not compatible with the eye and face protection equipment under consideration and should not be used with them.

To summarize the information provided on the compatibility of safety helmets and eye and face protection equipment, it is necessary to stress the importance of this issue for the safety and comfort of their users. Lack of proper fitting of this equipment may result not only in deterioration of its protective features, but also generate new threats, e.g. face injuries. For this reason, the compatibility of safety helmets

FIGURE 4.12 Checking the helmet's interaction with safety spectacles/goggles.

and eye and face protection should be taken into account both when purchasing equipment for workers and before first use at the worksite.

REFERENCES

Baszczyński, K. 2014. Simultaneous use of industrial safety helmets with other personal protective equipment. *Bezpieczeństwo Pracy – Nauka i Praktyka* 11(521):18–21.

Baszczyński, K. 2018. Effects of falling weight impact on industrial safety helmets used in conjunction with eye and face protection devices. *Int. J. Occup. Saf. Ergon.* 2(24):171–180.

Baszczyński, K., M. Jachowicz, and A. Jabłońska. 2011. Sprawozdanie z projektu 03.A.09. Opracowanie metody badania hełmów ochronnych w zakresie skuteczności ochrony przed uderzeniami poruszających się obiektów. [Project report 03.A.09. Development of the test method, for helmets, of effectiveness of protection against impacts of moving objects]. Warszawa: Centralny Instytut Ochrony Pracy – Państwowy Instytut Badawczy.

CEN [European Committee for Standardization]. 2001a. Personal eye-protection – non-optical test methods. EN 168:2001. Brussels: CEN.

CEN [European Committee for Standardization]. 2001b. Personal eye-protection – specifications. EN 166:2001. Brussels: CEN.

CEN [European Committee for Standardization]. 2006. Headforms for use in the testing of protective helmets. EN 960:2006. Brussels: CEN.

CEN [European Committee for Standardization]. 2012. Industrial safety helmets. EN 397:2012+A1:2012. Brussels: CEN.

5 Assessment of Protective Properties of Helmets and Eye Protectors

Central Institute for Labour Protection – National Research Institute

CONTENTS

5.1 ASSESSMENT OF TECHNICAL CONDITION OF PROTECTIVE HELMETS

5.1.1 BASIC METHODS OF LABORATORY TESTS OF PROTECTIVE HELMETS

Helmets play a very important role in protecting people in the working environment. For this reason, they must undergo appropriate laboratory tests to verify their protective parameters. Laboratory tests of helmets are carried out, for example, during the development of new constructions, selection of materials for their production, before they are placed on the market and during production quality control. These tests are conducted in specially prepared laboratories, according to standardized methods. These methods and requirements are most often derived from relevant standards, e.g. European Standards (EN) [CEN 2000a, 2000b, 2000c, 2000d, 2000e, 2008, 2012a, 2012b, 2012c, 2012d, 2012e, 2012f]. To illustrate the complexity of testing and evaluation of protective helmets used in industrial environments, the main test methods will be presented.

5.1.1.1 Shock Absorption

Shock absorption is one of the most important protective parameters of helmets designed for use in industrial conditions. This parameter shows what maximum force is transmitted by the helmet to the headform during an impact of a falling striker with assumed kinetic energy [Hulme 1996; CEN 2008, 2012a]. This shows how much energy can be absorbed by the helmet structure without exceeding the force that is considered dangerous for the user. The shock absorption of helmets intended mainly for protection against falling weight impact is determined by the "fixed headform" method [CEN 2012d]. This method consists in measuring the maximum value of the force with which the helmet acts on the headform during the impact of a falling striker. An example of a helmet shock absorption test stand is shown in Figure 5.1. This stand is placed on a monolithic base (1) of at least 500 kg which is designed to absorb the dynamic forces during impact on the tested helmet. A force transducer (2) with a headform (3) meeting the requirements of EN 960:2006 [CEN 2006] is mounted to the base. The tested helmet (4) is mounted on a headform according to the manufacturer's instructions. The helmet striker (5) may be of different shape and weight depending on the test methodology. Its face (striking) surface can be e.g. spherical, according to EN 397:2012+A1:2012 [CEN 2012a] or flat, according to EN 812:2012 [CEN 2012c]. The striker is attached to the trolley (7) which moves on vertical slideways (6). Thanks to the drive mechanism, the trolley with a striker can be pre-set at a certain height above the helmet,

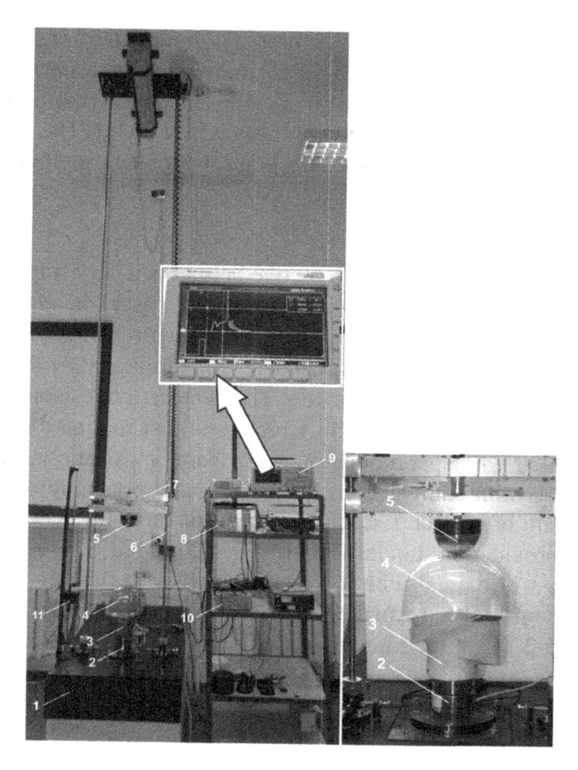

FIGURE 5.1 Test stand for shock absorption of protective helmets. Note: 1 – monolithic base, 2 – force transducer, 3 – headform, 4 – tested protective helmet, 5 – striker, 6 – striker trolley slideways, 7 – striker trolley, 8 – amplifier with analog filter, 9 – data acquisition system (oscilloscope with memory), 10 – control system for lifting, lowering and releasing the striker, 11 – gauge measuring final velocity of the striker.

thus obtaining the assumed kinetic energy of the striker when hitting the helmet. The mechanical part of the station is controlled by a control system (10).

The striker begins to fall when the electromagnetic latch in the trolley is released (7). Achievement of the target final velocity of the striker is verified using the gauge (11) at the last 2 cm of the fall. The force measuring apparatus consists of a force transducer (2), an analog amplifier with filter and a data acquisition system, e.g. digital oscilloscope (10). The amplitude and frequency characteristics of the apparatus are defined in relevant standards, e.g. EN 397:2012+A1:2012 [CEN 2012a] and EN 13087-2:2012 [CEN 2012d]. As a result of the test, the time course of the force under the headform is recorded and, after conducting numerical analysis, its maximum value is determined.

Prior to the shock absorption test, the helmets are preconditioned under different climatic conditions depending on the requirements, e.g. pursuant to EN 13087-1:2000 [CEN 2000a]. The necessity to conduct such tests results from the dependence of shock absorption on thermal conditions in which the helmets are used [Baszczyński 2014]. The most commonly used conditions are exposure to temperatures of −10°C, +20°C and +50°C, immersion in water, exposure to UV radiation (simulating accelerated ageing) or cycles of varying conditions, referred to as thermal shock, e.g. pursuant to EN 443:2008 [CEN 2008]. The time between the completion of preconditioning and the shock absorption test shall not exceed 1 min.

5.1.1.2 Resistance to Penetration

The resistance to penetration of the protective helmet is a very important parameter determining the extent to which it is able to protect the user's head from being hit by hard sharp-edged objects, e.g. falling pieces of ceramic tiles or sharp metal objects. Resistance to penetration of the helmet is determined mainly by the hardness of its shell and cushioning of the entire structure. The penetration resistance test is usually carried out according to the method in which a helmet mounted on a headform is hit by a sharp striker with specific parameters: shape, mass and velocity. The result of the test is a determination whether the sharp striker comes into contact with the headform as a result of impact. Examples of requirements for penetration resistance test methods are provided in standard EN 13087-3:2000 [CEN 2000b]. The tests may be carried out on a test stand with a structure similar to that shown in Figure 5.1, with the difference that a sharp striker is installed instead of a hemispherical one, as shown in Figure 5.2. Also, the force measuring equipment is not used on the stand.

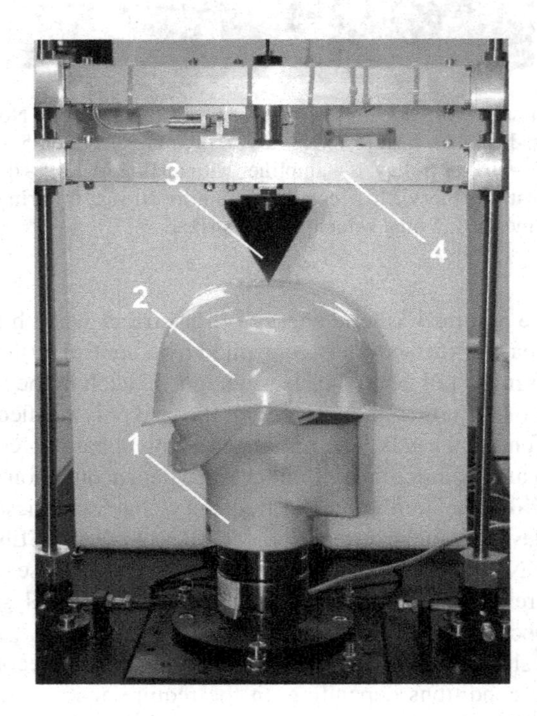

FIGURE 5.2 Stand for testing the resistance to penetration of protective helmets. Note: 1 – headform, 2 – tested protective helmet, 3 – striker, 4 – trolley for the striker.

As in the case of shock absorption, prior to the test of penetration resistance, helmets are preconditioned under different climatic conditions, depending on the requirements, e.g. in accordance with EN 13087-1:2000 [CEN 2000a]. The most common preconditions used are: temperatures of −10°C, +20°C and +50°C, immersion in water, UV exposure or cycles of varying conditions, e.g. pursuant to EN 443:2008 [CEN 2008]. The time between the completion of preconditioning and the shock absorption test should not exceed 1 min. Depending on the requirements, the helmets can be hit once or several times.

5.1.1.3 Resistance to Flame

In many workplaces, protective helmets are exposed to heat factors such as open flames. In the event of such a hazard, the helmet should protect the user's head from the effects of the flame and should not pose any additional risk itself, e.g. due to dripping of the shell material. In order to standardize the requirements in this respect, the requirements for different types of helmets and the testing methods contained, for example, in standard EN 13087-7:2000 [CEN 2000d], were specified. The idea of the test is to introduce the helmet into the burner flame for a given time and observe the effects. Some of the most important phenomena observed include the helmet catching fire, dripping of material, deformation of components, e.g. the shell, or the time of extinguishing the flame on the helmet after switching off the burner. An example of a stand for testing the helmets' resistance to flame is shown in Figure 5.3.

The stand includes a burner (1) with parameters defined e.g. in EN 13087-7:2000 [CEN 2000d], which is powered by pure propane. The tested helmet (3) is placed on a helmet stand (2) so that the surface of the shell, where it touches the end of the flame, is in a horizontal position. During the test, a 45 mm long flame is in contact with the shell for 10 s, and after extinguishing the flame, the duration of the flame on the shell is measured (if

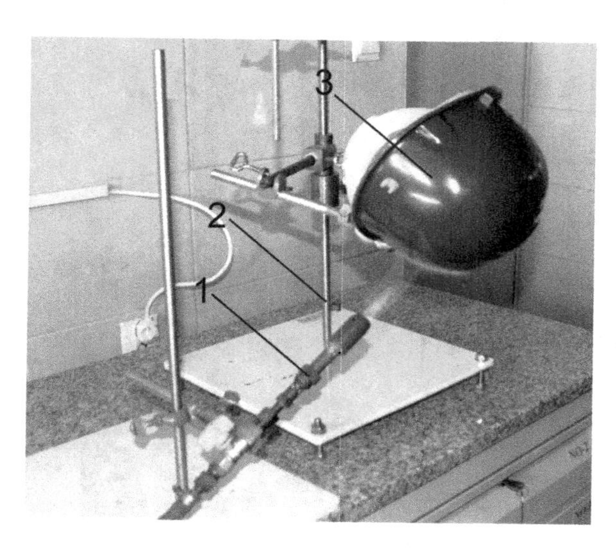

FIGURE 5.3 Stand for testing the helmets' resistance to flame. Note: 1 – propane burner, 2 – helmet stand, 3 – tested protective helmet.

this occurs). The presented testing method is used as a mandatory method for the evaluation of protective helmets used in industrial environments. For specialist helmets, e.g. for firefighters, more complex methods are used to create more difficult test conditions (e.g. higher flame temperatures, longer flame duration, sample preheating, etc.).

5.1.1.4 Breaking Strength of Chin Strap Anchorage

The chin strap is an accessory to the protective helmet. Its purpose is to prevent the helmet from falling off the head. An important feature of this piece of equipment is the value of the force at which the strap anchorages on the shell break off, thus freeing the user's chin. This protects the human head from the force exerted by the chin strap when the helmet peak is hit from below. The testing method for this parameter, as set out in EN 13087-5:2000 [CEN 2000c], consists in subjecting the chin strap to a slowly increasing force and measuring its maximum value at the moment of breaking the anchorages. An example of a test stand for the above-mentioned tests is presented in Figure 5.4.

The main element of the stand visible in Figure 5.4 is the testing machine (1) with which the test load, e.g. 20 N/min, according to EN 397:2012+A1:2012 [CEN 2012a] is set. In the working space of the testing machine there is a support structure (4) with a mounted headform (5). The headform is equipped with a tested protective

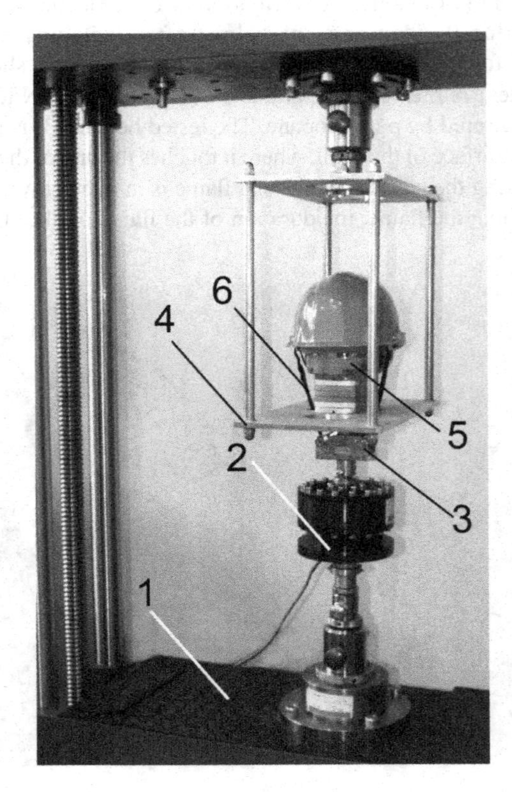

FIGURE 5.4 Stand for testing the strength of chin strap anchorages. Note: 1 – testing machine, 2 – force transducer, 3 – chin model, 4 – supporting structure, 5 – headform, 6 – chin strap.

helmet. The chin strap is passed through the chin model, (3) mounted on the measuring head (2). During upward movement of the crosshead of the testing machine, the belt and its anchorage points are loaded with linearly increasing force. This process continues until the anchorages break and the maximum force at which this occurs is determined. The result of the test shall be evaluated in relation to the requirements for a particular type of helmet, e.g. for helmets meeting the requirements of EN 397:2012+A1:2012 [CEN 2012a], the force shall range from 150 to 250 N.

5.1.1.5 Resistance to Lateral Deformation

In industrial conditions, there are workstations where the human head is exposed to lateral compressive forces, e.g. during handling operations. In such a situation it is necessary to use a protective helmet which protects against the compressive forces and does not become permanently deformed by them. For the assessment of protective parameters in this area, the test method set out in standards EN 443:2008 [CEN 2008] and EN 397:2012+A1:2012 [CEN 2012a] was developed. It consists in placing the helmet sideways between two parallel planes and subjecting it to compressive force. During the compression and relieving process, helmet deformation values are recorded. An example of a stand solution for conducting such tests is shown in Figure 5.5. The main element of this stand is a universal testing machine (1) equipped with a frame with parallel plates (2) that compress the helmet (4). The top plate (2) is moved along the guides of the test stand pressed by a tappet connected to the force

FIGURE 5.5 Stand for testing resistance to lateral deformation. Note: 1 – universal testing machine, 2 – horizontal compression plates, 3 – force transducer of the testing machine, 4 – tested helmet.

transducer (3) on the crosshead of the testing machine. The measuring cycle, i.e. loading at 100 N/min, maintaining a force of 430 N for 30 s, relieving to a force of 25 N and reloading to 30 N, is carried out by a testing machine equipped with a suitable control program. During the tests, the maximum helmet deflection and the permanent deflection after relieving the load are measured. In case of industrial helmets, the test is carried out as shown in Figure 5.5, and in case of specialized helmets, e.g. for firefighters, the test is carried out also in the forward and reverse direction.

5.1.1.6 Electrical Insulation

One of the dangers occurring at different workstations is the possibility of touching to metal parts that are electrically live with a helmeted head. This is a direct threat to human health and life, but it also poses a danger of causing an electrical short circuit through the helmet shell. For this reason, a number of helmet designs with electro-insulating properties were developed. Appropriate testing methods, such as those described in EN 13087-8:2000 [CEN 2000e], have been developed to verify these characteristics. They are based on the measurement of leakage current through the helmet to which 1200VAC is applied. In these methods, the test voltage is applied between:

- A point on the surface of a wet helmet shell and a metal headform to which it is attached,
- Two points on a dry helmet shell,
- Between an aqueous sodium chloride solution located inside and outside the helmet shell.

An example of a test according to the second method and the second measuring station is shown in Figure 5.6. The stand shown consists of an instrument (1) which

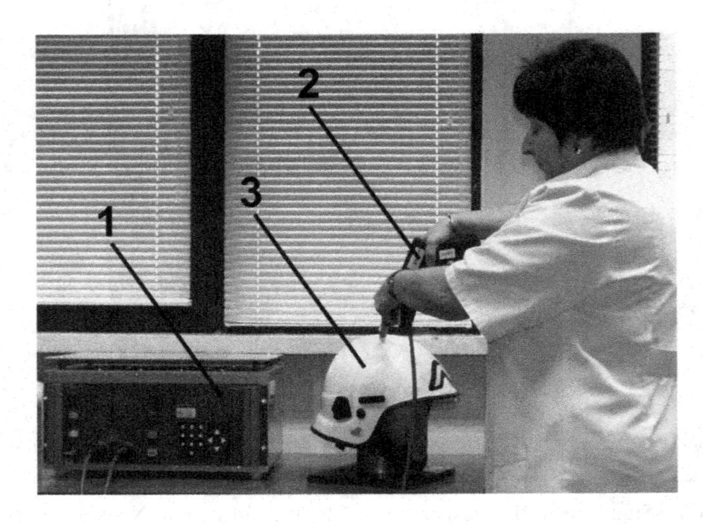

FIGURE 5.6 Stand for testing electrical insulation. Note: 1 – current leakage measuring device, 2 – measuring probes, 3 – tested helmet.

generates an increasing voltage at an assumed rate and measures the current flowing through the surface of the helmet shell (3). The voltage is applied using probes (2). In accordance with the requirements of EN 397:2012+A1:2012 [CEN 2012a] and EN 443:2008 [CEN 2008], the maximum leakage current shall not exceed 1.2 mA. In the case of specialized helmets, e.g. for workers carrying out live work, tests are conducted according to methods with higher test voltages.

5.1.2 Assessment of Safety Helmet Technical Condition by Their Users

Tests conducted so far on personal protective equipment and materials used for its production [Baszczyński 1992, 2008; Mewes 1998; Gijsman 1999] have shown changes in properties, including protective parameters, occurring over time. In the case of protective helmets which have been withdrawn after several years of use, studies have shown that the most important factors causing the loss of protective parameters included:

- Impacts, abrasions and other mechanical damage caused by contact with sharp and hard objects,
- Solar radiation, especially in the UV spectrum,
- Variable ambient temperatures,
- Molten metal splashes and open flames,
- Exposure to aggressive chemicals,
- Self-degradation process of materials (plastics) used in the production of helmets.

The effects of these factors may be so severe that the helmets no longer perform their protective function. Therefore, it is necessary to exercise appropriate supervision over the technical condition of protective equipment during its entire lifetime. The duty of such supervision rests mainly with the employers who have equipped their employees with such equipment, but also with the users themselves. Such oversight is effective when it involves carrying out controls on two levels:

- Immediately before each use (assessment carried out directly by an appropriately trained user),
- Periodically (e.g. once a year), performed by a competent, specially prepared person within the company or directly by the manufacturer (e.g. their authorized service centre).

The first of these levels is extremely important as it allows to detect the damage causing loss of protective parameters and thus minimizes the probability of using faulty equipment. Furthermore, it is performed by a person whose health and, in extreme cases, life depends on the technical condition of this equipment. The second level requires more knowledge and experience, but guarantees the detection of more complex damage cases.

The assessment of the usefulness of the protective helmet for further use should begin by checking that the period during which the manufacturer guarantees its protective properties has not expired. This inspection should be carried out on the basis of information provided by the manufacturer, e.g. the date of manufacture permanently marked on the shell and a record in the instructions for use. If the period indicated by the manufacturer is found to have expired, the helmet should be taken out of service, irrespective of its appearance.

Performing an independent technical condition assessment of the protective helmet requires the user to have adequate knowledge. The basic information in this area is provided by the manufacturers' instructions for use. Unfortunately, they speak of potential damage in general terms and do not provide concrete examples, and therefore the evaluator is often unable to assess whether the observed change is a significant damage or not. Tests carried out in CIOP-PIB in the years 2008–2010 [Baszczyński et al. 2008] allowed development of materials which present, among other things, typical damage to protective helmets that support independent inspection of personal protective equipment during use. The presented material was based on the results obtained.

5.1.2.1 Assessment of Protective Helmet Shells Technical Condition

Checking the technical condition of the helmet shell is a very important element in assessing its suitability for use. The main task of the shell is to take over the impact of a dangerous object, partially absorb its energy and transfer the remaining part to the helmet's harness. The shell also prevents direct contact of the user's head with a dangerous object.

The shells are made of various types of plastics, such as polyethylene, ABS plastic or polyester-glass laminates, which undergo aging processes, especially under the influence of sunlight [Baszczyński 1992; Gijsman 1999; Mewes 1998]. One of the main effects of ageing plastics is an increase in their rigidity and brittleness. In the case of helmets, this effect may manifest itself in the appearance of characteristic silent crackles during slight compression. To check the helmet in this regard, gently squeeze it with your hands in the direction shown in Figure 5.7 and listen for the sound accompanying the increase in pressure. Shell cracking may indicate degradation of the material and the presence of microcracks in the shell, and thus the loss of flexibility and durability.

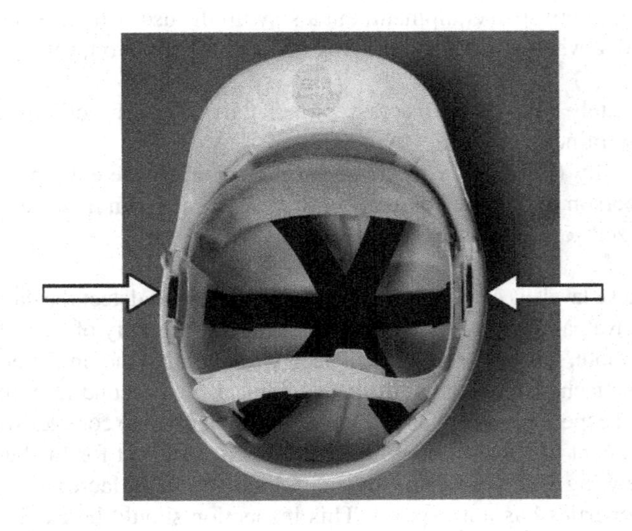

FIGURE 5.7 Testing the crackle of the helmet shell.

Loss of protective parameters of the helmet may also be indicated by easily noticeable damage to the shell. The most important include:

- Cracks on the surface and throughout the thickness of the shell,
- Deep visually noticeable deformations,
- Large area depigmentation and discoloration,
- Deep abrasions over large areas,
- Large surface chips in the shell material or damage resulting in sharp edges,
- Edge chips causing sharp edges.

Examples of damage for which the helmet should be retired are shown in Figure 5.8.

If the helmet shell on the inside of the helmet is fitted with a cushioning lining, it should also be inspected. In case of such damage as carpet separation from the shell, cracking or crushing of carpet fragments, colour change, etc. the helmet must be retired.

5.1.2.2 Assessment of Harness Technical Condition

The harness is one of the most important parts of the helmet for the absorption of impact energy of a moving object. Its damage, due to a strong impact on the helmet, can lead to the shell coming into contact with the user's head and transferring practically all the impact energy to it.

The purpose of the assessment of the technical condition of the harness is to check its damages and the reliability of the connection with the helmet shell. The simplest way to test is to apply pressure in the direction shown in Figure 5.9. You can exert this force by pressing your fist on the harness when the helmet is pointing downwards.

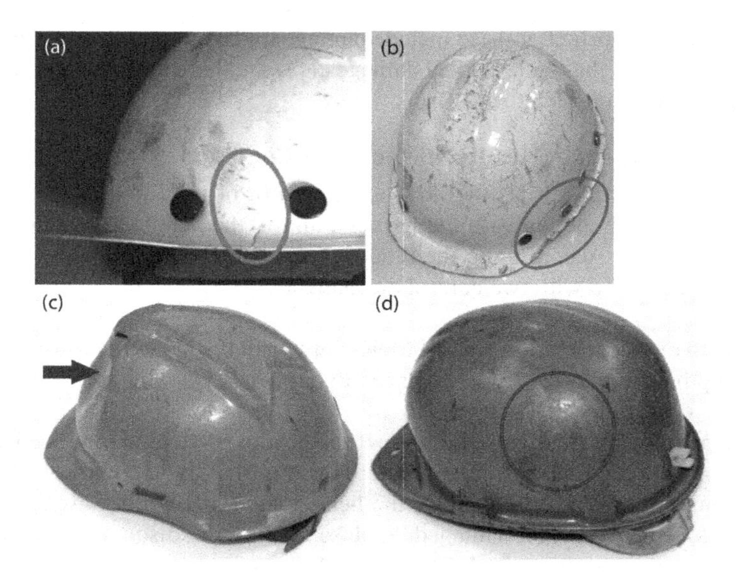

FIGURE 5.8 Examples of helmet shell damage: (a) helmet shell cracks, (b) helmet edge chips, (c) helmet shell deformation, (d) large shell surface damage.

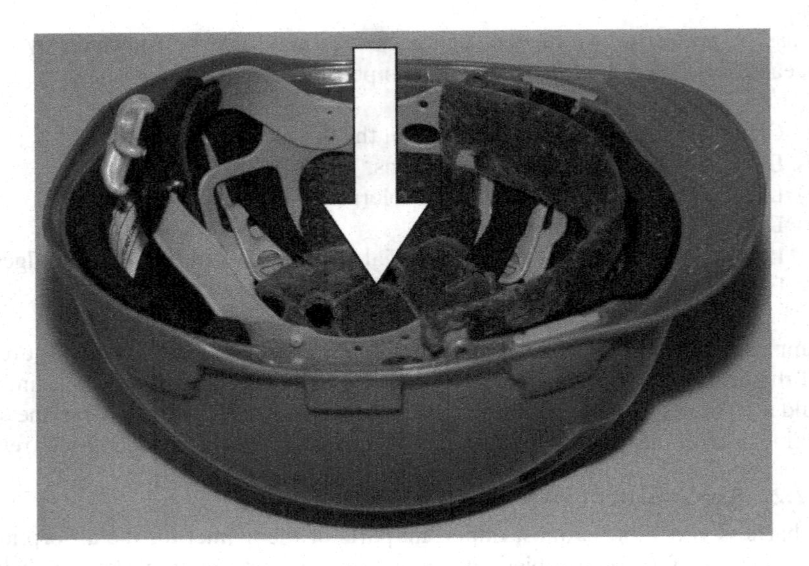

FIGURE 5.9 Checking the helmet's harness behaviour when exerting pressure.

During this examination, it must be observed that the shell does not separate from the harness and that the bands are not broken. The appearance of such effects indicates the need to retire the helmet.

The technical condition of the harness and the durability of its connection to the helmet shell are also evidenced by other changes that can be easily noticed by the user. The most important of these include damage to the attachments of the harness and their sockets in the shell. Critical damage also includes cuts in the harness straps and rips in their seams (Figure 5.10).

5.1.2.3 Assessment of Headband Technical Condition

The headband, which girdles the head at the height of the forehead and the base of the skull, in conjunction with the harness, allows the helmet to be held steadily on the user's head. According to the requirements of EN 397:2012+A1:2012 [CEN 2012a], it is equipped with two control mechanisms:

- Depth of placement in the shell (height at which the helmet is worn),
- Length (fitting to the circumference of your head).

When inspecting the technical condition of the headband, hold it with one hand and the edge of the shell with the other. Then, by gently shaking the helmet, check that the headband does not separate from the helmet shell and that the wearing height adjustment elements are not damaged. Such separation is a disqualifying effect and requires the helmet to be retired. By alternately stretching and squeezing the headband slightly (Figure 5.11) check that the set length does not change. If this happens, the helmet should not be used.

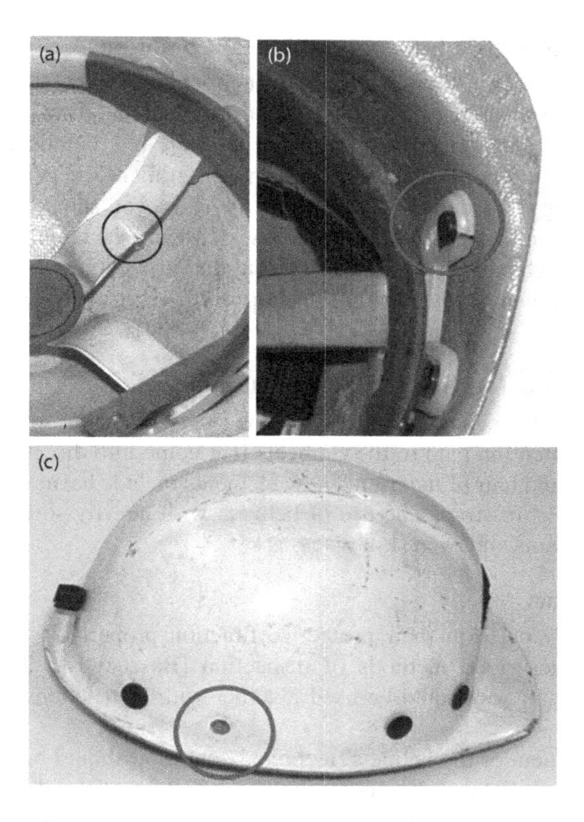

FIGURE 5.10 Damage to the harness. (a) cut harness straps, (b) rupture of harness attachment to the shell, (c) no rivet for harness attachment.

FIGURE 5.11 Checking the durability of the headband length setting.

5.1.2.4 Assessment of Chin Strap Technical Condition

The main task of the chin strap is to keep the helmet on the head when bending over and making rapid movements, e.g. when arresting a fall. The strap performs its function properly if its anchorages to the shell/harness are not damaged and the length adjuster does not automatically change the length setting of the strap.

When checking the technical condition of the strap, you should examine its connections with the helmet shell/harness and assess whether its length does not change by stretching it slightly with your hand. During the inspection, attention should also be paid to the mechanical damage to the strap itself, i.e. to the presence of cuts and abrasions.

5.1.2.5 Assessment of Helmet Hygienic Condition

When examining the technical condition of the safety helmet, also check its soiling, with particular attention paid to the elements that come into direct contact with the skin of the head and hair of the user. These elements include harness, headband with a sweatband and chin strap. The use of helmets with heavily soiled elements may cause skin irritations and even skin diseases.

5.1.2.6 Summary

Helmets can only perform their protective function properly if they are in good condition. The described methods of inspection [Baszczyński 2010] and examples of typical damages should be used together with instructions provided by the manufacturers.

Visual assessment of the helmets' technical condition should be made each time before their use, while other manual checks (i.e. crackles of the shell, connection of the harness and the headband to the shell, stability of the length setting of the headband and chin strap) should be done according to the intensity and conditions of use, but at least once a year. In addition, regardless of the result of the technical condition check, the principle should be applied that a protective helmet that has been heavily impacted should be retired – whether the damage is noticeable or not.

5.2 EYE AND FACE PROTECTORS

5.2.1 Laboratory Tests on Eye and Face Protectors

The designs of eye and face protectors described in Subsection 5.2.2 shall be tested to see whether the properties of the materials used for the protective structures, as well as their complete construction, meet the requirements. A set of characteristics that should be observed in the materials and structures of personal protective equipment is described in CEN [2001a, pr. ISO 16321-1]. As with the way the division of eye and face protectors is defined, the characteristics relating to these products may have slightly different definitions depending on the geographical area in which the standard is issued. However, this does not change the fact that both the test methods and requirements for all types of eye and face protective equipment can be divided into two main groups. They are: tests of optical and non-optical properties (mechanical parameters). European standards EN 166:2001 [CEN 2001a], EN 167:

2001 [CEN 2001b] and EN 168:2001 [CEN 2001c], which divide the requirements and test methods for eye and face protectors into optical and non-optical, are an example of this approach.

At the time of writing this monograph, work on the development of new international standards in this area was underway. The European standards quoted above will be replaced by international standards. EN 166:2001 [CEN 2001a], which sets out the requirements for eye and face protectors, will be replaced by ISO 16321-1 [pr. ISO 16321-1] and EN 167:2001 [CEN 2001a] (specifies optical test methods) and EN 168:2001 [CEN 2001b] (specifies non-optical test methods) by ISO 18526-2 [pr. ISO 18526-2] (specifies the reference test methods for determining the physical optical properties) and ISO 18526-3 [pr. ISO 18526-3] (specifies the reference test methods for determining the physical mechanical properties), respectively. New standards containing the description of methods for testing physical optical and mechanical properties are complemented by ISO 18526-1 [pr. ISO 18526-1] (specifies geometrical optical properties) and ISO 18526-4 [pr. ISO 18526-4] (contains the description of headforms used for testing optical and non-optical parameters).

Considering that with the date of application of the new international standards, eye and face protectors whose protective properties have been assessed on the basis of the results of laboratory tests carried out in accordance with the European standards will not cease to be used, both the requirements of European standards (EN) and the new international standards (ISO) are referred to below.

The basic idea of evaluating optical and mechanical properties presented in the European and international standards cited above has remained unchanged.

The most important element in the assessment of optical properties for all eye and face protectors are the properties that determine the transmission of optical radiation through the filter or protective lens (oculars). In both European and international standards, these properties are determined on the basis of the measured spectral transmittance characteristics of the tested transparent element. The changes to be noted when comparing the requirements for optical properties in the European and international standard include, among others:

- Taking into account for the calculation of parameters determining the colour recognition behaviour when looking through filters the spectral distribution of radiation emitted from LED sources,
- Extension of the spectral range used in the evaluation of infrared radiation protection filters from 2000 to 3000 nm, and
- A change in the approach to assessing the refractive properties of filters and protective glasses without filtering action.

ISO 18526-2 [pr. ISO 18526-2] also contains a detailed description of how to carry out an evaluation for automatic welding filters.

In mechanical properties tests, the most important element for evaluation is the resistance of the material (filter, protective lens or complete protection) to impact. Changes in the scope of mechanical properties tests consist, among others, in the introduction of new categories of mechanical resistance.

In the new international standards, the issues related to the estimation of measurement uncertainties are discussed in a much more elaborate way, which is very helpful in the interpretation of results obtained from laboratory tests.

5.2.1.1 Physical Optical Properties

Tests of optical properties concern such features which allow one to determine the characteristics of protective lenses (both unmounted filters and protective lenses without a filter effect, as well as complete protection). Optical properties tests assess the quality of workmanship, parameters related to refraction, transmission and reflection of optical radiation and the field of vision. In order to assess the quality of the workmanship, it is necessary to determine the parameters which allow evaluation of e.g. the degree of surface undulation, pits, inclusions, scratches or other defects in the material of which the protective lens is made. Verification of the quality of workmanship shall be carried out by visual evaluation as well as by determining the indicator/parameter – reduced luminance factor, which determines the amount of light scattered on the sample. As the number of defects increases, the amount of light scattered on the tested protective lens will increase. If the protective lens does not have a corrective function, its optical power (parameters determining refractive capacity) must be at zero or very low levels. According to the EN 166:2001 [CEN 2001a] standard, for protective glass panes without corrective effect, which can be classified in the so-called first optical class, the spherical power must not exceed ±0.06 diopters and the astigmatic (cylindrical) power must be less or equal to 0.06 diopters. The definition of optical class is omitted in ISO 16321 [pr. ISO 16321-1] although it does contain requirements for optical power. The values of the parameters determining the spherical and astigmatic (cylindrical) power are determined for welding filters (passive and active) and other types of eye and face protectors (eye shields, face shields as well as spectacles and goggles). The maximum spherical power values for welding filters and other eye and face protectors are ±0.06 diopters and ±0.12 diopters, respectively. The astigmatic (cylindrical) power for welding filters and other eye and face protectors shall be less than or equal to 0.06 diopters and 0.12 diopters, respectively. However, ISO 16321-1 [pr. ISO 16321-1] specifies an additional requirement in this respect for products with so-called enhanced optical performance. The maximum optical power values for products with enhanced optical performance specified in ISO 16321-1 [pr. ISO 16321-1] shall comply with the requirements for first optical class specified in EN 166:2001. As a consequence of this change, users who have used eye and face protectors classified in accordance with EN 166:2001 [CEN 2001a] to the first optical class must – pursuant to ISO 16321-1 [pr. ISO 16321-1] – use eye protectors with enhanced optical performance.

The permissible deviations of the optical power values, as described in EN ISO 8980-1:2017-11 [ISO 2017], are strictly defined for the corrective lenses.

In the case of filters, the most important element of their evaluation is to determine the level of transmission and reflection of optical radiation that falls on them. The principle for the definition of coefficients for the assessment of the transmission and reflection of optical radiation is described in detail in Chapter 2 (Basic construction of eye and face protectors).

Measurements of the spectral transmittance and reflectance characteristics from which the coefficients used to assess the filters are determined are made

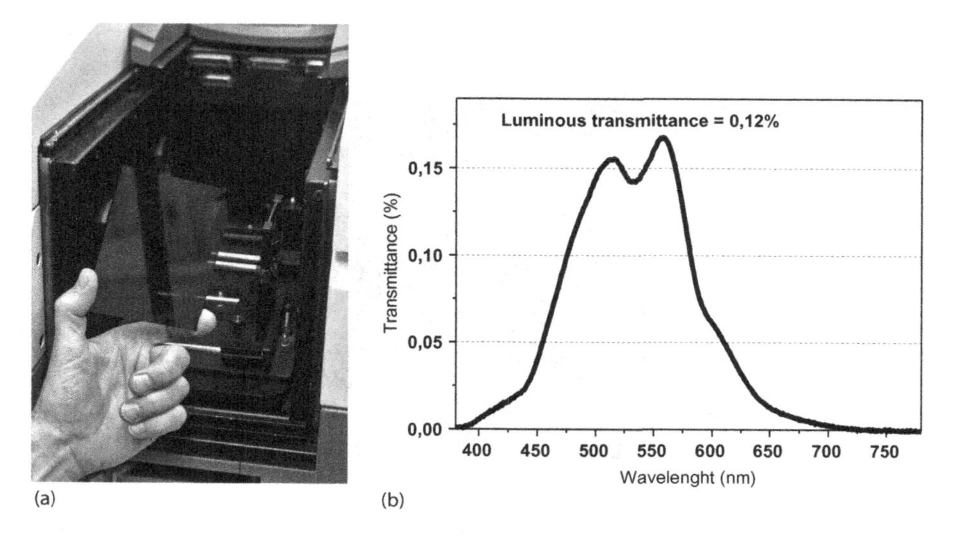

FIGURE 5.12 (a) Photograph of the measuring chamber of the spectrophotometer used in CIOP-PIB [CIOP-PIB], (b) Spectral transmittance characteristics of a typical (designation 12) welding filter.

using optical spectrophotometers. Such devices must have a high measuring resolution and sensitivity. This is due to the very high spectral optical density of some types of optical protective filters. For example, the level of visible radiation transmission for typical filters used in arc welding is at a thousandth of a percent. Figure 5.12 shows a photograph of the measuring chamber of the spectrophotometer used in CIOP-PIB and the spectral transmittance characteristics of a typical welding filter.

For the determination of many optical properties of eye and face protectors, a highly specialized measuring and testing equipment is used that is not commercially available. An example of such a device is an apparatus for measuring residual optical powers at the level of hundredths of diopters. Measurements of optical power at this level are not possible with typical commercially available dioptrimeters. Figure 5.13 shows a photograph of an apparatus for measuring optical powers of the eye and face protectors used in CIOP-PIB.

Some optical properties of protective lenses are determined for new samples and samples conditioned under the conditions specified in the standard [CEN 2001b, 2001c, pr. ISO 18526-2, pr. ISO 18526-3]. This is a typical assessment procedure called "before/after". The conditions in which protective lenses are conditioned include ultraviolet radiation, high and low temperatures and abrasion. The optical properties of the protective lenses can also be determined in real time during the conditioning process or during the operation of selected external factors.

For anti-fog safety glasses, a fogging time is determined. It is determined by measuring the changes in the optical transmittance factor that occur directly as a result of the vapour deposition on the tested protective lens.

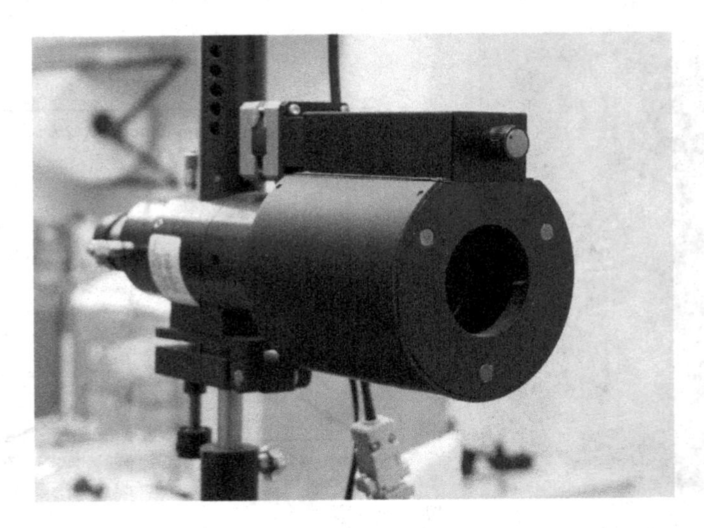

FIGURE 5.13 Apparatus for measurement of optical powers for eye and face protectors [CIOP-PIB].

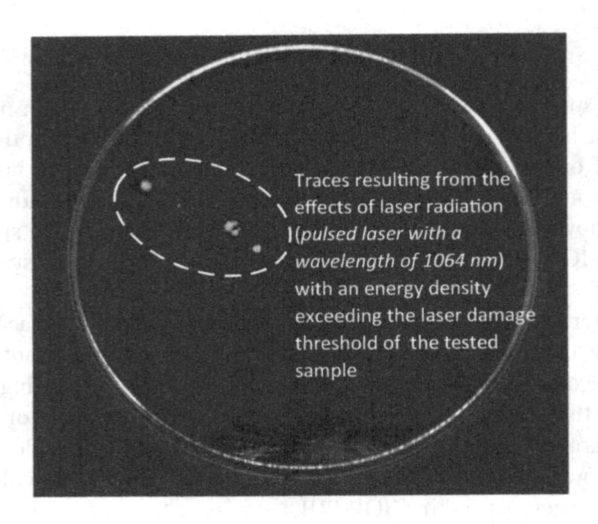

FIGURE 5.14 Photograph of traces resulting from the effects of laser radiation (pulsed laser with a wavelength of 1064 nm) with an energy density exceeding the laser damage threshold of the tested sample [CIOP-PIB].

With regard to filters protecting against laser radiation, energy resistance measurements are performed [CEN 2017]. They involve measurements of the spectral transmittance factor when the laser radiation is applied to the tested filter. Figure 5.14 shows a photograph of traces resulting from the effects of laser radiation (pulsed laser with a wavelength of 1064 nm) with an energy density exceeding the laser damage threshold of the tested sample.

5.2.1.2 Physical and Mechanical Properties

Non-optical tests are performed on selected elements of frames and housing and the complete construction of the eye and face protector. These tests consist in determining the mechanical strength parameters of new or conditioned eye and face protectors under specific conditions [CEN 2001c, pr. ISO 18526-2]. Conditioning can take place at high/low temperatures, in humidity (including immersion in water) and under conditions of exposure to optical radiation (ultraviolet and infrared). Non-optical testing also includes examining whether protective parameters are maintained as a result of exposure factors such as molten metals and hot solids, liquid streams, dust and fumes, chemicals and the electric arc.

The basic parameter for assessing the mechanical strength of eye and face protectors – assessed in accordance with the requirements of European standard EN 166:2001 [CEN 2001a] – is resistance to high-speed particle impact. This test shall be carried out according to the method described in EN 168:2001 [CEN 2001c]. It consists in hitting the protective lens and the side parts of the housing or frame with a 6 mm steel ball. Under the conditions described in the standard, the ball can strike the protective lens at speeds of 45, 120 or 190 m/s. Mechanical resistance is determined according to the applied speed of the ball. Low mechanical resistance (F symbol) indicates ball speed of 45 m/s, medium (B symbol) – 120 m/s and high (A symbol) – 190 m/s. For goggles, the classification includes velocities of 45 or 120 m/s and, in the case of face shields, all of the mentioned speeds (45, 120 or 190 m/s). The safety spectacles impact test shall only be carried out at the lowest ball speed (45 m/s). This does not mean that the spectacles materials would be damaged when hit with a ball at a higher speed. In the case of spectacles made of, for example, polycarbonate, a 120 or even 190 m/s ball impact test would most likely have a positive result. The spectacles only protect the eyes and a small area around them. If we compare the simulation carried out using a ball with the hazards caused by eye and face exposure to real splinters, their effects are comparable. It is also easy to predict that splinters with higher energy pose a real threat not only to the eyes, but also to the rest of the face. It is then necessary to use a protective lens whose surface covers a larger eye area (goggles) or the whole face (face shield). Therefore, the conditions under which laboratory tests on the mechanical strength of particular types of eye and face protectors (spectacles, goggles, face shields) are carried out correspond to the risks that occur in the actual working environment. It would be inappropriate to provide safety spectacles that have been certified as resistant to 120 or 190 m/s ball impact. Under exposure to splinters whose effect is comparable to that of a 120 or 190 m/s ball impact, goggles or face shields are required. Figure 5.15 shows a photograph of a part of the apparatus for testing the resistance of eye and face protectors to high-speed particles used by CIOP-PIB.

The new international standard ISO 18526-3 [pr. ISO 18526-3] has extended the range of mechanical resistance tests and changed the speed values of the steel ball in the high-speed particle impact test. The components that interact with the eye protectors tested can be steel balls of different weights and diameters or a 500 g cone. These elements hit the tested protection from different heights and at different speeds. The masses and diameters of the balls depend on the test method adopted. According to the new international standards [ISO 18526-3], there are three test methods for ball impact: static load, drop ball and ballistic testing.

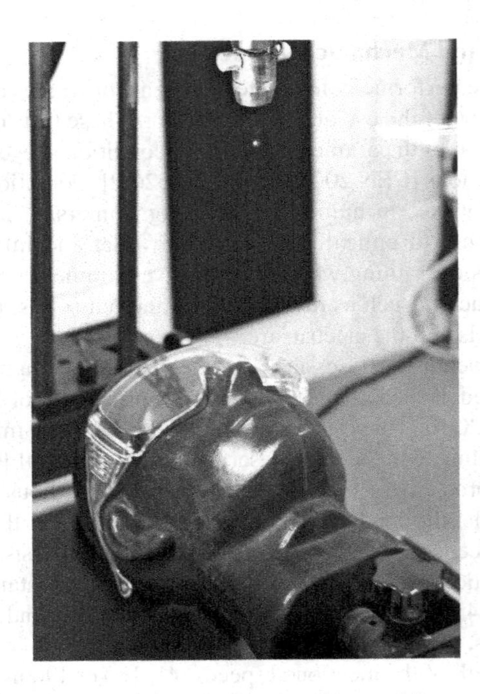

FIGURE 5.15 Apparatus for testing the resistance of eye and face protectors to high-speed particles [CIOP-PIB].

In the static load test, a 22 mm diameter steel ball presses the test sample with a force of 100 N for 10 s. As a result of this test, the mechanical resistance category is defined as minimum robustness.

The drop ball test uses balls with diameters of 16 and 22 and 25.4 mm and corresponding masses of 16 and 43 and 66.8 g. In both cases, the balls hit the test sample from a height of 1.27 m. Regardless of the mass of the ball, the speed at which it hits the test sample is about 5 m/s (the mean value of the standard gravity at $g = 9.81$ m/s^2 is assumed). As a result of this test, the mechanical performance level category is determined as: level 1 (for a ball of 16 g) and level 2 (for a ball of 44 g) and the so-called *Basic Impact* for a ball of 66.8 g.

For ballistic tests, as for the high-speed particle impact tests described above, according to EN 168:2001 [CEN 2001b], a ball with a diameter of 6 mm and a mass of 0.86 g shall be used. However, the speed values of the ball have been changed. The test ball may impact the sample at the following speeds: 45, 80 and 120 m/s. As a result of this test, the mechanical resistance category is determined as: impact level C (for ball speed of 45 m/s), impact level D (for ball speed of 80 m/s) and impact level E (for ball speed of 120 m/s).

When hitting the tested protection, the cone has a mass of 500 g and hits the sample from a height of 1.27 m. As a result of this test, a category of mechanical performance known as *High Mass* is determined.

Depending on the type of protection, the mechanical performance level is determined. For sunglasses this may be minimum robustness or impact levels 1, 2 or 3. For protection for occupational use these can be *Basic Impact*, *High Mass* or C, D or E impact levels. For welding filters, minimum resistance and *High Mass* are specified. Impact level 2 and *High Mass* are determined for sports eye protection. The new requirements also include categories of mechanical resistance to squash ball impact. There are two categories: *Squash* and *Racquetball and Squash 57*. For testing, 40 m/s impact balls with diameters of 40 and 57.3 mm and corresponding masses of 24 and 99.2 g are used.

5.2.1.3 Features of Tested Eye and Face Protectors

The methodology for laboratory testing of eye and face protectors refers directly to the characteristics that they should possess. Two approaches to the way in which the characteristics of personal protective equipment under examination are determined are discussed below. The approach presented in European standard EN 166: 2001 [CEN 2001a] and new international standard ISO 16321-1 [pr. ISO 16321-1] is discussed. Table 5.1 shows the features of eye and face protectors with the division of tests into optical and non-optical, as described above, according to European standard EN 166:2001 [CEN 2001a].

Regardless of the type and application, all personal protective equipment should be tested to meet the basic requirements. Another set of features depends on the area of application of the protection (particular requirements). The attributes that eye and face protection products should have in order for their use to benefit the user (optional requirements) are also specified. This approach to examining the features of a particular eye and face protector in laboratory tests does not mean that a feature of any kind can be examined in laboratory tests (from the list in Table 5.1). For example, a feature such as increased infrared reflectance is only tested for infrared protection filters. This means that for a particular type of personal protective equipment only the set of features that can be attributed to it should be selected.

The new international standard ISO 16321-1 [pr. ISO 16321-1] explicitly mentions the relationship between the required features of certain types of personal protective equipment. In this standard, the features examined are assigned to specific types of eye and face protectors, such as:

- Lenses without deliberate filter action,
- Ultraviolet filters (UV filters),
- Infrared filters (IR filters),
- Sunglare filters for occupational use,
- Welding filters,
- Filters used in glass blowing,
- Cover plates,
- Complete protectors without deliberate filter action,
- Complete protectors with UV, IR and sunglare filters,
- Complete welding protectors,
- Complete protectors for glass blowing,
- Mesh protectors.

TABLE 5.1
Summary of Eye and Face Protector Features According to EN 166:2001

No.	Tested Feature	Type of Test (Optical/Non-optical)
Basic Requirements		
All eye protectors should meet the basic requirements		
1	General design	Non-optical
2	Materials	
3	Headband	
4	Fields of vision	Optical
5	Astigmatic and prismatic spherical powers	
6	Luminous transmittance	
7	Scattering of light	
8	Material and surface quality	
9	Minimum robustness	Non-optical
10	Increased robustness	
11	Resistance to ageing	
12	Resistance to corrosion	
13	Resistance to ignition	
Particular Requirements		
Depending on the intended use		
14	Protection against optical radiation	Optical
15	Protection against high-speed particles	Non-optical
16	Protection against molten metals and hot solids	
17	Protection against droplets of liquids	
18	Protection against large dust particles	
19	Protection against gas and fine dust particles	
20	Protection against short circuit electric arc	
21	Lateral protection	Optical
Optional Requirements		
Additional requirements apply to the characteristics of eye protectors that may be considered beneficial to the user during use		
22	Resistance to surface damage by fine particles	Non-optical
23	Resistance to fogging	Optical
24	Enhanced infrared reflectance	
25	Protection against high-speed particles at extreme temperatures	Non-optical

Source: CEN [European Committee for Standardization], CEN/TC 85, Eye protective equipment. Personal eye-protection –Specifications EN 166:2001, CEN, Brussels, 2001a.

The list of features for which requirements are specified in the international standard ISO 16321-1 [pr. ISO 16321-1] has been extended in relation to the European standard EN 166:2001 [CEN 2001a]. These include, among others:

- Chemical resistance,
- Reflection from mesh protectors,

- Penetration of vents and slots,
- Arc detection sensitivity of automatic welding filters,
- Ability to see through close-up lenses,
- Uniformity,
- Anti-reflective coating reflectance.

Thus, the new international standard ISO 16321-1 [pr. ISO 16321-1], taking into account also the elements not previously considered, indicates in an unambiguous way a set of features that certain types of personal protective equipment should have. It also indicates the corresponding requirements and laboratory test methods.

5.2.2 Assessment of Eye and Face Protectors' Technical Condition

5.2.2.1 Causes of Eye and Face Protectors Degradation at the Working Environment

As with other personal protective equipment, both the materials used and the production technology determine the quality and, in particular, the protective properties of eye and face protectors. Production quality supervision by the manufacturer and competent and authorized testing laboratories are also important. Testing of the features of individual protective products is carried out in accordance with the described principles of technical condition assessment, according to the applicable regulations. Testing laboratories evaluate mainly the characteristics of new, unused products. While the test procedures for eye and face protectors shall take into account selected aspects of use by conditioning the tested products prior to testing, these conditions merely simulate the factors to which the protectors, including eye and face protectors, may be exposed under actual use conditions.

When examining the issue of assessing the technical condition of eye and face protectors, it should therefore be borne in mind that the actual conditions in the working environment may differ from those under which new products were tested. Users themselves should be aware of the fact that the protection they use should be checked as often as possible to ensure that the product in use has not been damaged, which could adversely affect its protective properties or user comfort. With regard to eye and face protectors, i.e. products where the most important element is the one we look through, it is also extremely important to ensure that all the transparent elements are clean. Both scratched and dirty safety lenses, filters and visors significantly impair visual comfort, which also has a direct impact on work safety. The proper quality control of eye and face protectors cannot be performed without the necessary knowledge of the harmful and hazardous factors and their influence on individual components of the structure. Awareness of such factors helps to properly assess the technical condition of eye and face protectors.

Eye and face protectors are a barrier to harmful and hazardous factors and are therefore damaged themselves. These factors include, but are not limited to, chips of solid substances and impacts, liquid splashes, natural and artificial optical radiation, other atmospheric conditions, high-voltage hazards, as well as the process of slow self-degradation. When analysing the causes of damage to eye and face protectors, the way in which they are used cannot be overlooked either. This includes both the intended use as well as appropriate maintenance, cleaning and disinfection.

Chips of solids and impacts cause scratches and cracks. An extremely difficult parameter to estimate is that which, under conditions of actual use, would indicate a correlation between factors such as chips of solids and impacts and the mechanical strength – including resistance to scratches – of the components of the eye and face protectors. At the outset, it was noted that the features of the new products are tested under laboratory conditions. When working in conditions where there is a risk of chipping of solids or impacts, it is necessary, in order to take due care of the safety of the performed work, to check systematically that these factors have not caused damage in the form of cracks or scratches. This applies in particular to the transparent elements (protective lenses, visors or filters) that are directly in the line of sight. The damage is mainly caused by splashes of those liquids which react with the materials of which the protection is made.

The factors stimulating such a reaction are elevated temperature and the composition and chemical properties of the liquid, and usually both at the same time. In the case of transparent elements, splashes of liquid cause e.g. tarnishing of the protective lens or filter and point damage to the so-called pitting. Then, in practice, the user has significantly limited visibility and discomfort of further work. In the case of other structural elements, such as headgears or frames and housing, splashes of aggressive chemicals can reduce their mechanical strength. When working under exposure to splashes of liquids, it is necessary to check that there are no areas on the surface of the equipment used that indicate contact of the chemical with the material of the protection applied (pits, matt stains, colour change).

Optical radiation is a very important element that can damage eye and face protectors. Firstly, the solar radiation [Corti 2010] should be mentioned, especially the ultraviolet radiation range (in the solar radiation spectrum there is mainly UV-B and UV-A range). Ultraviolet radiation is one of the main factors influencing the degradation (photodegradation process) of many plastics [Premamoy 2010], which are also used in the construction of eye and face protectors. Intense solar radiation also leads to heating of materials to relatively high temperatures (above 70°C), which can cause a significant change in the elastic properties (i.e. elasticity) of plastic components, and in extreme cases even their melting – plastic deformation [Godovsky 1992]. Frames made of plastic and exposed to intense sunlight will not be able to hold the lens or filter in the right position. Visible radiation – excluding laser radiation – even if it is very intense (e.g. visible radiation emitted during welding processes), does not contribute to the degradation of the materials used for the manufacture of eye and face protectors. In the case of artificial infrared radiation (e.g. metallurgical processes) there is exposure to relatively high temperatures (even above 1000°C – thermal ageing), which may cause a significant change in physical and mechanical properties and destruction of polymer materials [Pecora 2016].

A special case of optical damage to the eye and face protective equipment is the damage caused by laser radiation. Laser radiation, due to the possibility of obtaining very high energy in a small area, can, in extreme cases, cause perforation or inflammation of the materials from which the eye and face protectors against this very radiation are made. It should be stressed that the most dangerous effects of harmful optical radiation on the polymer materials from which eye and face protectors are made should include those that are invisible to the "naked eye" and cause a decrease in strength properties. When working under exposure to optical radiation, it is necessary to monitor that the applied eye and face protectors have not deteriorated in the form of impairment of the structure of plastic materials and tarnishing of the protective lens or filter and, in extreme cases, melting or charring.

Other factors that weaken the protective properties of eye and face protectors include the negative influence of weather conditions, mainly the effects of variable temperatures (cold/heat), temperature limits and humidity. Using eye and face protectors at relatively low temperatures can lead to a significant reduction in the elasticity of the plastics of which the protectors are made, and thus to damage to their structure (mainly cracks). High temperature conditions may result in the melting of materials (exceeding the glass transition temperature – T_g in which the material shows high stability of mechanical and thermal properties). Rain or humidity is particularly dangerous for commonly manufactured welding shields made of materials such as pressboard or fibres. Drastic changes in temperature from above to below accelerate the impairment of strength properties. Changing temperature and humidity is also a threat to the increasingly used electronic components/modules integrated with eye and face protectors. When working under conditions of exposure to variable temperatures or humidity, it is important to ensure that there are no alarming changes in the structure of the materials of which the protective equipment is made, such as cracks, surface damage, etc. In the case of eye and face protectors with built-in electronic components, special attention must be paid to the proper functioning of these components.

High-voltage evoked hazards are mainly related to situations where a spark jump can occur due to a significant potential difference. The energy of the spark can lead to inflammation and, in extreme cases, even charring of the face shield material. When working under high-voltage conditions, care must be taken to ensure that the eye protectors used are not damaged in a way that indicates the loss of insulation. This applies both to safety glasses and visors, as well as to the entire structure which may contain metal or other conductive elements.

Self-destruction of materials is a very broad issue. The term "self-destructive" indicates that the destruction process takes place regardless of whether or not the intensified factors described above (e.g. ultraviolet radiation, humidity, temperature, etc.) occur. What time it takes for materials to self-destruct depends on how and for how long they are stored, used and maintained. Materials made of certain plastics (especially rubbers, i.e. elastomers from aliphatic polymer chains) are subject to the process of oxidation (oxi-degradation) which results in their "natural" degradation [Yashchuk 2012]. Oxidation causes brittleness of materials as a result of the appearance of microcracks that appear at relatively low loads [Odegard 2011]. It is therefore extremely important to monitor the use and storage time of eye and face protectors. This applies primarily to those protectors which contain elements made of materials susceptible to oxidation [Jansen 2002]. The moment when damage caused by self-destruction of materials may occur can be unpredictable and surprising. The surface or structure of some elastomers affected by degradation may appear to be intact. It is only when they are used that such materials decompose uncontrollably, which in practice manifests itself in a lack of proper fit (elastomer elements) as well as brittleness under stress. Therefore, a principle to be followed as good practice for the use of all types of eye and face protectors should be adopted: proper storage and marketing control of protective equipment and mandatory control of the technical condition of all components made of materials susceptible to self-destruction before each use.

With regard to the equipment in question, eye and face protectors, damage caused by misuse and inadequate maintenance may also occur. This is mainly due to the lack of appropriate care for the protective equipment entrusted to the employee. In the case of eye and face protectors, the damage falling within this group results, among other things, from the lack of knowledge of the strength characteristics of the materials from which they are made. The most common example is the improper placement of safety spectacles in such a way that the lens or filters come into direct contact with the surface.

Materials used for lenses and protective filters are not always covered with curing layers to protect against scratches. These are often materials with high mechanical resistance to impact, but with limited abrasion resistance. Frequent contact of such a material with another, especially rough, substrate material significantly shortens the protector's life span. Also, the use of inappropriate procedures for cleaning transparent elements is common. Improper maintenance and cleaning can also cause significant loss of protective and functional properties. This applies both to dirty protective lenses and filters, as well as to the soiling of the entire structure of the protective device. If, for example, vents in the protective goggles, which are tightly fitted to the face, are clogged, they immediately lose their required ventilation properties and no longer perform their function.

When discussing the use and maintenance of eye and face protectors, attention should also be paid to the electronic components integrated into the protectors. Battery or accumulator-powered electronic modules can be damaged, e.g. by installing unsuitable, damaged batteries with electrolyte leakage. Improper use of eye and face protectors with implemented electronic components may result in, among other things, damage to the screens (breakage, scratches) or loss of contact (vision) which clearly precludes further use. Another good practice for the use of eye and face protection products is therefore their proper, prescribed use, as well as the provision of appropriate storage conditions and maintenance, cleaning and, if required, disinfection.

The technical inspection of safety spectacles, goggles and face shields should consist of two stages: assessment of the condition of the transparent elements (filter or protective lens) and assessment of the fixing of these elements (frame, housing or headgear). The assessment is subjective in nature and consists of an organoleptic examination to determine whether damage, mainly mechanical, has occurred, both to the translucent and other elements of the protective device.

Defects of transparent elements include scratches, tarnishes and cracks, which translates into the quality of vision and, consequently, into the safety of the eyes and comfort of visual work. The most frequent damage to the fastening elements includes cracks, deformations, excessive loosening of the temple length and angle adjustment elements, which in turn leads to impairment in the fitting of eye protectors to the face (the ability of spectacles, goggles and face shields to remain in place).

The following describes and illustrates the method used to assess the technical condition of typical elements of eye and face protectors, such as protective lenses and filters, frames of safety spectacles, goggles and face shields, automatic welding filters and welding protectors (welding shields and helmets, welding goggles). The examples described and illustrated show the most frequent damages that result from the above-mentioned factors.

5.2.2.2 Assessment of the Technical Condition of Transparent Elements

In order to perform the title evaluation, a thorough visual inspection of the protective lens and visor or filter must be carried out for surface and structural damage over the entire thickness of the material. Damage to the optical elements is better visible when observed against a dark background: it is recommended to check the technical condition on a dark surface. Identifying damage such as cracks, scratches, tarnishing or pits within the field of vision shall be considered as a negative result. Examples of damage to the protective lens in the form of pits and scratches are shown in Figure 5.16.

5.2.2.3 Assessment of the Technical Condition of the Safety Spectacles Housing

Assessment of the technical condition of the safety spectacles housing should begin with a thorough visual inspection for cracks, deformations, excessive loosening of the elements for adjusting the length and angle of the temple, the ability of the spectacles to remain on the head and other damage. Identifying damage such as rupture of the temple or side shield, deformation of the temple or side shield, loosening of the elements for adjusting the length and angle of the temple, resulting in impairment of the ability to remain on the head, shall be considered as a negative result. Examples of components to be checked when assessing the technical condition of safety spectacles frames are shown in Figure 5.17.

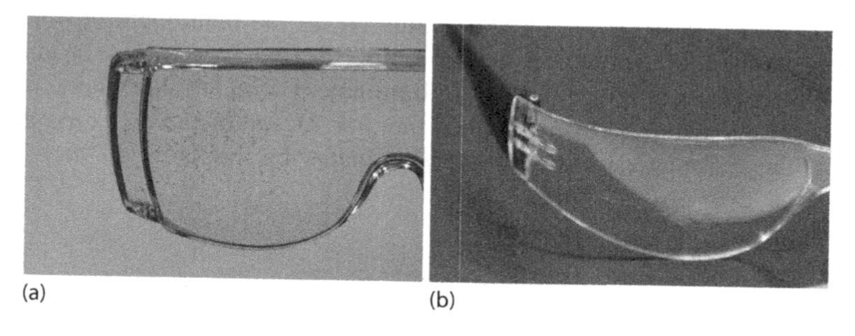

(a) (b)

FIGURE 5.16 Damage to safety lenses in the form of pitting (a) and scratches (b).

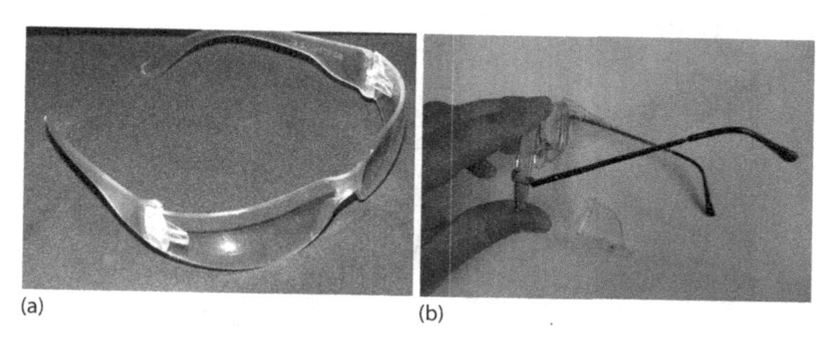

(a) (b)

FIGURE 5.17 Elements inspected during technical condition assessment. (a) Checking the connection of the temple with the protective lens. (b) Checking the connection of the spectacles housing to the temple.

5.2.2.4 Assessment of the Technical Condition of Goggles and Face Shields

In order to properly assess the technical condition of goggles and face shields, a thorough visual inspection of the protective lens and the housing/headgear should be carried out with respect to scratches, tarnishes, cracks, deformations, the ability of the goggles and face shields to remain on the head, soiling of the elements directly in contact with the user's head and the ventilation system's patency (applies to goggles). Identification of damages such as scratches or tarnishing of the surface of protective lenses within the field of vision; any cracks and deformations of the protective lenses or their frame, housing or headgear; heavy soiling of elements coming into direct contact with the user's head or obstruction of the ventilation system (applies to goggles) shall be considered a negative result. Figure 5.18 shows an example of obstructed (clogged) goggle vents.

5.2.2.5 Assessment of the Technical Condition of Automatic Welding Filters

Inspection of the technical condition of automatic welding filters (which automatically darken as a result of the initiation of a welding arc) installed in helmets should include assessment of the condition of the filter (cracks, discoloration of the screen, etc.) and verification of filter activation without the initiation of a welding arc. If any test ends with a negative result or a change resembling the damage presented is noticed, it may indicate a loss of protective properties of the automatic welding filter. In such a case, the filter must be withdrawn from use and returned for re-examination and evaluation to the manufacturer, its service centre or another competent person. The simplest way to verify that the automatic welding filter is working properly is to illuminate it with a lighter flame or direct infrared radiation emitted from a typical radio or TV remote control onto the filter detector. The automatic welding filter, illuminated by the flame of the lighter or

FIGURE 5.18 Obstructed (clogged) goggle vents.

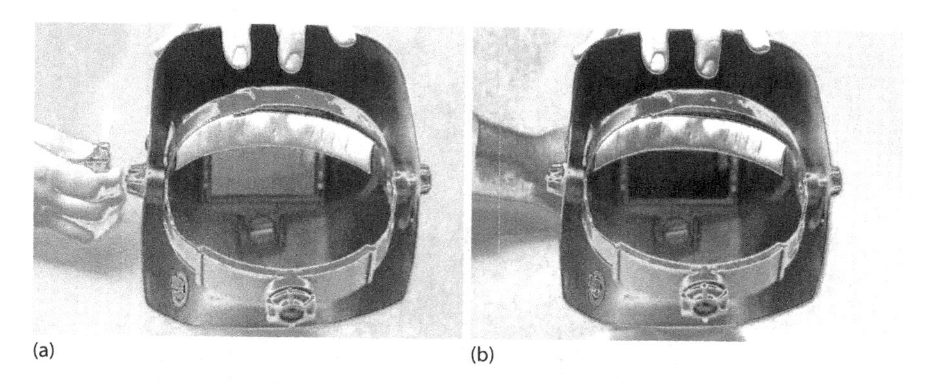

(a) (b)

FIGURE 5.19 Checking the activation of the automatic welding filter (filter mounted on the welding helmet): (a) light state and (b) dark state.

exposed to the radiation emitted by the remote control, should darken. The lack of darkening indicates a high probability of filter damage. Figure 5.19 shows how to check whether the automatic welding filter is working with a lighter.

5.2.2.6 Assessment of the Technical Condition of Welding Shields and Safety Helmets

Inspection of the technical condition of welding shields and helmets, including fitted filters, should include assessment of the condition of the filter and protective lens and the fixing of these elements in the frame, evaluation of the condition of mounting the hand grip (applies to shields) and the headgear (applies to helmets) to the shield or helmet and visual assessment to determine whether any damage to the shield's body has occurred, such as deformation, loss of lightproofness, excessive moisture in the material, etc. If any assessment ends with a negative result or a change resembling the damage presented is noticed, it may indicate a loss of protective properties of the welding shield or helmet. In such a case, the welding shield/helmet must be withdrawn from use and returned for re-examination and evaluation to the manufacturer, its service centre or another competent person. Figure 5.20 shows the damage to the weld shield body joints which results in the loss of its lightproofness. Such a shield does not provide full protection against glare caused by the welding arc and UV radiation.

5.2.2.7 Assessment of the Technical Condition of Welding Goggles

The inspection of the technical condition of the welding goggles with filters fitted should include an assessment of the condition of the filter and protective lens and the attachment of these elements to the frame; an assessment of damage to the frame, such as deformation, loss of lightproofness, etc.; testing the correct functioning of the hinged element (with welding filters fitted) and testing the quality of the headband and protective cushions in direct contact with the user's face. If any of the assessments ends with a negative result or a change resembling the damage presented is

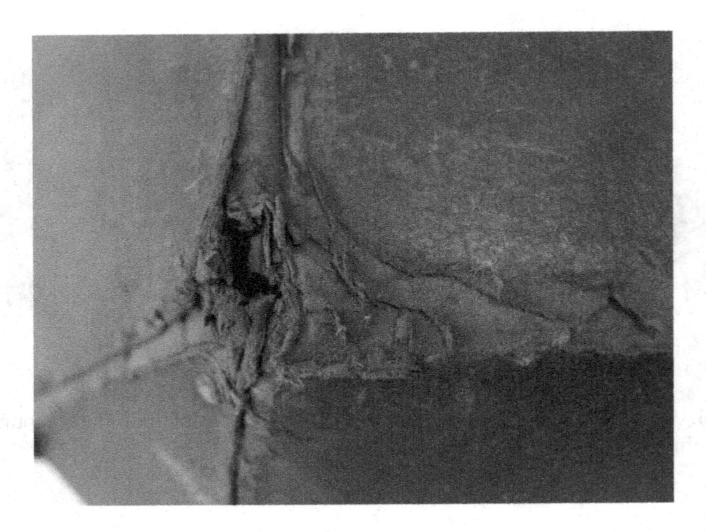

FIGURE 5.20 Damage to joints resulting in loss of lightproofness (UV and glare protection is not provided).

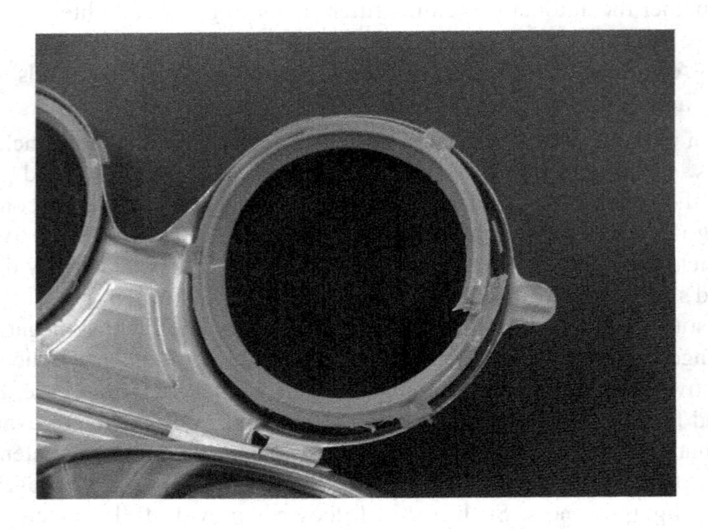

FIGURE 5.21 Damage to the ring used to fix the welding filter in the goggle housing.

noticed, it may indicate a loss of protective properties by the goggles. In such a case, they must be withdrawn from use and returned for re-examination and evaluation to the manufacturer, its service centre or another competent person. Figure 5.21 shows typical damage to the ring used to attach the welding filter to the goggle housing.

The assessment of the technical condition of all eye and face protective equipment should be carried out by a trained worker and under the supervision of persons responsible for health and safety at the workplace. Quality control is an important element in the system of managing personal protective equipment.

REFERENCES

Baszczyński, K. 2014. The effect of temperature on the capability of industrial protective helmets to absorb impact energy. *Eng. Fail. Anal.* 46:1–8.

Baszczyński, K., M. Dolecki, and G. Łaszkiewicz. 1992. Sprawozdanie z projektu 238/V – Nowelizacja Polskiej Normy PN 86/Z-08110 Przemysłowe hełmy ochronne – w zakresie określenia dopuszczalnego czasu użytkowania i magazynowania oraz odporności termicznej hełmów ochronnych [Project report 238/V. Amendment to the Polish Standard PN 86/Z-08110 Industrial safety helmets – In terms of determining the allowable time of use and storage and thermal resistance of industrial protective helmets]. Warszawa: Centralny Instytut Ochrony Pracy – Państwowy Instytut Badawczy.

Baszczyński, K., and M. Jachowicz. 2010. Wytyczne do samodzielnej oceny stanu technicznego przemysłowych hełmów ochronnych [Guidelines for self-assessment of the technical condition of industrial safety helmets]. *Bezpieczeństwo Pracy – Nauka i Praktyka* 2(461):10–12.

Baszczyński, K., M. Jachowicz, A. Jabłońska et al. 2008. Opracowanie materiałów informacyjnych wspomagających pracowników w samodzielnej kontroli użytkowanych środków ochrony indywidualnej. [Project report 3.S.17. Development of the guide to support employees in individual control of personal protective equipment]. Warszawa: Centralny Instytut Ochrony Pracy – Państwowy Instytut Badawczy.

CEN [European Committee for Standardization]. 2000a. Protective helmets. Test methods. Conditions and conditioning. EN 13087-1:2000. Brussels: CEN.

CEN [European Committee for Standardization]. 2000b. Protective helmets. Test methods. Resistance to penetration. EN 13087-3:2000. Brussels: CEN.

CEN [European Committee for Standardization]. 2000c. Protective helmets. Test methods. Retention system strength. EN 13087-5:2000. Brussels: CEN.

CEN [European Committee for Standardization]. 2000d. Protective helmets. Test methods. Flame resistance. EN 13087-7:2000. Brussels: CEN.

CEN [European Committee for Standardization]. 2000e. Protective helmets. Test methods. Electrical properties. EN 13087-8:2000. Brussels: CEN.

CEN [European Committee for Standardization], CEN/TC 85. 2001a. Eye protective equipment. Personal eye-protection –Specifications EN 166:2001. Brussels: CEN.

CEN [European Committee for Standardization], CEN/TC 85. 2001b. Eye protective equipment, Personal eye-protection –Optical test methods EN 167:2001. Brussels: CEN.

CEN [European Committee for Standardization], CEN/TC 85. 2001c. Eye protective equipment, Personal eye-protection – Non-optical test methods, EN 168:2001. Brussels: CEN.

CEN [European Committee for Standardization]. 2006. Headforms for use in the testing of protective helmets. EN 960:2006. Brussels: CEN.

CEN [European Committee for Standardization]. 2008. Helmets for firefighting in buildings and other structures EN 443:2008. Brussels: CEN.

CEN [European Committee for Standardization]. 2012a. Industrial safety helmets. EN 397:2012+A1:2012. Brussels: CEN.

CEN [European Committee for Standardization]. 2012b. High performance industrial helmets EN 14052:2012+A1:2012. Brussels: CEN.

CEN [European Committee for Standardization]. 2012c. Industrial bump caps EN 812:2012. Brussels: CEN.

CEN [European Committee for Standardization]. 2012d. Protective helmets. Test methods. Shock absorption. EN 13087-2:2012. Brussels: CEN.

CEN [European Committee for Standardization]. 2012e. Protective helmets. Test methods. Field of vision. EN 13087-6:2012. Brussels: CEN.

CEN [European Committee for Standardization]. 2012f. Protective helmets. Test methods. Retention system effectiveness. EN 13087-4:2012. Brussels: CEN.

CEN [European Committee for Standardization]. 2017. Personal eye-protection equipment. Filters and eye-protectors against laser radiation (laser eye-protectors) EN 207: 2017-07. Brussels, Belgium: CEN.

Corti, A., S. Muniyasamy, M. Vitali, S. H. Imam, and E. Chiellini. 2010. Oxidation and biodegradation of polyethylene films containing pro-oxidant additives: Synergistic effects of sunlight exposure, thermal ageing and fungal biodegradation. *Polym. Degrad. Stabil.* 95:1106–1114.

Gijsman P., G. Meijers, and G. Vitarelli. 1999. Comparison of the UV-degradation chemistry of polypropylene, polyethylene, poliamide 6 and polybutylene terephthalate. *Polym. Degrad. Stabil.* 65:433–441.

Godovsky Y. K. 1992. *Thermophysical Properties of Polymers*, Berlin-Heidelberg-New York: Springer-Verlag.

Hulme, A. J., and N. J. Mills, 1996. The performance of industrial helmets under impact. An assessment of the British standard BS 5240 PT. 1, 1987. School of Metallurgy and Materials. University of Birmingham. Health & Safety Executive contract research report no. 91/1996.

ISO [International Organization for Standardization], ISO/TC 172/SC 7, Ophthalmic optics and instruments, Geneva, Switzerland (2017), EN ISO 8980-1: 2017-11 clause 5.2.2. Uncut finished spectacle lenses – part 1: Requirements for monofocal and multifocal lenses.

Jansen, J. A. 2002. Characterization of Plastics in Failure Analysis. *ASM Handbook. Vol. 11: Failure Analysis and Prevention.* W. T. Becker, and R. J. Shipley, eds. Materials Park, OH: ASM International. 11:437–459.

Mewes, D. 1998. Ageing of components, technical work equipment and personal protective equipment made of plastic. BIA-Info 5/98. *Arbeit und Gesundheit spezial.* 5:20.

Odegard, G. M., and A. Bandyopadhyay. 2011. Physical aging of epoxy polymers and their composites. *J. Polym. Sci. Part B Polym. Phys.* 49:1695–1716. doi: 10.1002/polb.22384.

Pecora, M., Y. Pannier, M.-C. Lafarie-Frenot, M. Gigliotti, and C. Guigon. 2016. Effect of thermo-oxidation on the failure properties of an epoxy resin. *Polym. Test.* 52:209–217.

Premamoy, G. 2010. *Polymer Science and Technology, Plastics, Rubber, Blends and Composites.* New Delhi: McGraw Hill Education.

Project ISO 16321-1. Eye and face protection for occupational use – Part 1: General requirements.

Project ISO 18526-1. Eye and face protection – Test methods – Part 1: Geometrical optical properties.

Project ISO 18526-2. Eye and face protection – Test methods – Part 2: Physical optical properties.

Project ISO 18526-3. Eye and face protection – Test methods – Part 3: Physical and mechanical properties

Project ISO 18526-4. Eye and face protection – Test methods – Part 4: Headforms.

Yashchuk O., F. S. Portillo, and E. B. Hermida. 2012. Degradation of polyethylene film samples containing oxo-degradable additives. *Procedia Mater. Sci.* 1:439-445. doi: 10.1016/j.mspro.2012.06.059.

6 From the Authors

Technological progress in each of the areas of the broadly understood work environment results in the fact that humans are no longer necessary for the direct execution of many technical or professional activities as they were in the past. Consequently, safety during work with automated processing and production lines is usually ensured by the use of collective protective equipment and industrial automation systems that automatically react to potential hazards. The direct participation of a human in operating this type of systems is often limited to the proverbial "pushing the button" and safe observation of the process in the hazard-free zone. This case confirms one of the most important principles of the safe work organization, which is to ensure human safety by applying appropriate technical and organizational solutions.

The last element of protecting human health and life in the work environment is personal protective equipment. Its application results from the fact that we try to reduce the risk associated with work and all non-professional activities to the highest possible degree, because we are not able to surround ourselves with some kind of barrier preventing the entry of hazardous factors into the closest human environment.

There are, and probably will always be, situations where the only way to ensure the safety of our eyes and head is to use personal protective equipment in the form of industrial safety helmets and different types of spectacles, goggles or face shields. Therefore, one can risk a statement that head protectors as well as eye and face protectors will never go out of use and that the constructions of such protectors will continue to be developed.

The growing awareness of the hazards and consequences of not using personal protective equipment, including protective helmets and eye and face protectors, has a significant contribution in reducing the number of personal accidents. The role of occupational health and safety publications has always been to spread knowledge about safety in the work environment. The intention of the authors of this monograph was to share theoretical knowledge, and what is more important, practical knowledge in the scope of protective helmets and eye and face protectors. The presented issues show that, for the proper use of the protective equipment described in the monograph, it is important to have a comprehensive knowledge of the materials used to create the protective equipment, the features of complete constructions, methods of laboratory assessment, the way of selection to the existing hazards while preserving compatibility, ending with a responsible decision to withdraw from service the used protective equipment due to the lost protective properties. The goals set by the authors concerning broadening and updating the knowledge of recipients of head protectors – protective helmets, eye and face protectors – due to technological progress and new trends in design, have been achieved. However, the final assessment is up to the reader.

6 From the Authors

Index

Note: Page numbers in bold and italics refer to tables and figures, respectively.

S

T

U

V

W

Printed in the United States
by Baker & Taylor Publisher Services